D0984468

Toxins and Plant Pathogenesis

Toxins and Plant Pathogenesis

Edited by

J. M. DALY

Department of Agricultural Biochemistry
University of Nebraska

B. J. DEVERALL

Department of Plant Pathology
and Agricultural Entomology
The University of Sydney

ACADEMIC PRESS

A Subsidiary of Harcourt Brace Jovanovich, Publishers

Sydney New York London
Paris San Diego San Francisco São Paulo Tokyo Toronto
1983

ACADEMIC PRESS AUSTRALIA
Centrecourt, 25–27 Paul Street North
North Ryde, N.S.W. 2113

United States Edition published by
ACADEMIC PRESS INC.
111 Fifth Avenue
New York, New York 10003

United Kingdom Edition published by
ACADEMIC PRESS, INC. (LONDON) LTD.
24/28 Oval Road, London NW1 7DX

Printed in Australia

National Library of Australia Cataloguing-in-Publication Data

---

Toxins and plant pathogenesis.

Bibliography.
ISBN 0 12 200780 8.

1. Micro-organisms, Phytopathogenic.
2. Microbial toxins. I. Daly, J. M.
(Joseph Michael), date. II. Deverall, Brian J.
(Brian James).

581.2'3

Library of Congress Catalog Card Number: 83-70711

Academic Press Rapid Manuscript Reproduction

Contents

3

Molecular Modes of Action

D. G. GILCHRIST

4

Roles of Toxins in Pathogenesis

SYOYO NISHIMURA AND KEISUKE KOHMOTO

5

Future Prospects in Toxin Research

RICHARD D. DURBIN

Contributors

Numbers in parentheses indicate the pages on which the authors' contributions begin.

RICHARD D. DURBIN (159), Plant Disease Resistance Research Unit, Department of Plant Pathology, 1630 Linden Drive, University of Wisconsin, Madison, Wisconsin 53706, U.S.A.

D. G. GILCHRIST (81), Department of Plant Pathology, University of California, Davis, California 95616, U.S.A.

KEISUKE KOHMOTO (137), Faculty of Agriculture, Tottori University, Tottori, Japan

V. MACKO (41), Boyce Thompson Institute for Plant Research, Cornell University, Ithaca, New York 14853, U.S.A.

SYOYO NISHIMURA (137), Nagoya University, Faculty of Agriculture, Plant Pathology Laboratory, Chikusu-ku, Nagoya 464, Japan

ROBERT P. SCHEFFER (1), Department of Botany and Plant Pathology, Michigan State University, East Lansing, Michigan 48824, U.S.A.

Preface

In the last decade, there has been a notable surge of interest in microbial toxins of potential importance in plant disease. The toxin concept has had advocates since the inception of plant pathology as a science in the latter half of the 19th century, but research on the topic has followed a bumpy path, strewn with the intellectual rocks of scepticism and contention. The climate of opinion, however, began to change perceptibly in the late 1940s. This change is rooted in several intertwined scientific developments.

In the last two decades, there have been a number of well-documented reports of host-specific, or host-selective, toxins as integral factors in disease development. One of these instances, Southern leaf blight of corn, involved disastrous epidemics in the USA in 1970 and 1971. These epidemics provided the sharp economic jolt that encouraged funding for research on toxins. Further, this example engaged the attention of geneticists and molecular biologists internationally, because of its potential significance in understanding male sterility in plants.

Simultaneously with these biological developments, there was the extremely rapid development, both in theory and practice, in separation and purification of natural products; for example, by gas and high pressure liquid chromatography. Further, improvements in diagnostic techniques, such as mass and nuclear magnetic resonance spectroscopy, provided opportunities for elucidation of structure on milligram quantities of toxins in a matter of weeks or even less. No longer are gram quantities essential for chemical elucidation. It is important to note that, in addition to establishing structural details, proper use of these techniques can establish purity of toxin. The question of purity has been a stumbling block in the acceptance of much of the earlier work on the role of toxins in disease.

It is not surprising that the first successes with these new techniques were recorded for several non-specific toxins produced by plant pathogens. In addition, the discovery that one of these, fusicoccin, had growth regulant properties stimulated the interest of plant physiologists throughout the world. Subsequently, the structural determination of the fungal toxin, tentoxin, and several bacterial toxins permitted sophisticated studies of the mode of action which demonstrated that it is possible to interpret their biological effects in terms of modern biochemistry. Knowledge of the chemistry of the host-specific toxins developed later, but advances in just the last three years have been remarkable. In 1975, the structure of only one host-specific toxin was known. From 1979 to the present, seven additional toxins have had structural assignments that appear to be sound. These accomplishments will provide a solid foundation for research leading to a better understanding of the chemical basis for plant disease resistance and susceptibility.

This topic was selected as the theme for a symposium for the 4th International Congress of Plant Pathology held in Melbourne, Australia, in August 1983. The authors were asked to prepare essays on the current status of research on toxins. The book does not attempt an exhaustive review of the literature. Rather, the essays represent the authors' judgement of progress and problems in certain aspects, with particular emphasis on more recent, and in some cases unpublished, findings. Some readers will note that host-selective toxins appear to have received emphasis. This should be expected and welcomed as it seems clear that host-selective toxins play a significant role in plant disease. The role extends beyond pathogenesis into epidemiology and plant breeding and because of this, international interest in them is high.

The book is intended for research workers, university teachers and advanced students in plant pathology, plant biochemistry, botany and plant breeding.

We should like to thank the contributors for co-operating with our timetable and Academic Press Australia for their helpful involvement in the venture. We also thank Anne-Louise Deakin for her help in preparing the index.

1

Toxins as Chemical Determinants of Plant Disease

ROBERT P. SCHEFFER

I. The Significance of Toxins That Affect Plants

The purpose of this chapter is to give the reader a general background and a broad perspective of toxins in plant pathology. I will indicate the toxins that are considered to be significant, summarize why they are important to our understanding of plant disease, and in general terms summarize how we arrived at our current state of understanding. Exciting recent developments and the all-important details in the research will be left for later chapters.

We do not have a widely-accepted definition of "toxin", so my first task is to define the word as it will be used in this volume. Toxins in plant pathology have three essential features: a) toxins are products of microbial pathogens; b) toxins cause obvious damage to plant tissues; and c) toxins are known with confidence to be involved in disease development. Feature (c) is included because fungi and bacteria in culture produce many substances that are toxic to plant cells, but which have no known roles in disease development. Enzymes released from pathogens usually are rather arbitrarily excluded from the toxin category; thus, most toxins known to date in plant pathology have low molecular weights. Terms such as "vivotoxin" and "pathotoxin" are frequently encountered in the literature, but these terms are confusing and hardly needed if the definition of toxin given

TOXINS AND PLANT PATHOGENESIS
ISBN 0 12 200780 8

above is followed. Even the word "phytotoxin" can be misleading because it is often used to indicate a toxic substance from higher plants (Harper & Balke, 1981).

Toxins have significance in agriculture, a fact that is not fully appreciated. The destructive effects of a number of infectious microorganisms can be traced to action of their toxins. The southern leaf blight of maize is perhaps the most dramatic example; this disease had an alarming impact on our economy in 1970-1971 (Ullstrup, 1972). An earlier example was the blight disease of oats, which in 1946-1948 caused vast losses in North America (Meehan, 1950). Toxins produced by the pathogens (fungi of genus *Helminthosporium*) were major factors in the destructive process of these two diseases. Other examples include toxins produced by bacteria of the genus *Pseudomonas*, which are involved in diseases of several major crops (soybean, bean, tobacco, stone fruits, and citrus). There are other good examples, but these are adequate as economic justification for study of plant toxins. This important aspect of plant science deserves far more attention than it has received in the past. There are hopeful signs of corrective attention.

We are primarily concerned in this volume with toxins as a scientific problem. A knowledge of toxins can contribute to our understanding of plant diseases at both the molecular and ecological levels, as I will show in the following discussions. These studies can in turn be placed in still broader perspective, if we compare the roles of toxins, allelopathic compounds, pheromones, antibiotics, and other substances that may be involved in inter- and intra-species interactions. All these substances belong in the field of chemical ecology, an exciting and rapidly-developing area of biology.

An understanding of disease development and disease resistance in plants is a major, long-term goal. To date, the most definitive progress toward this goal has been made in studies on toxins and on tumor-inducing DNA(T-DNA) from *Agrobacterium tumefaciens* (Kahl & Schull, 1982). The T-DNA model appears to be applicable to relatively few plant diseases; the toxin model, on the other hand, may apply to a much larger group. We know that certain toxins are keys to an understanding of pathogenicity, virulence, and host selectivity for several plant-infecting microorganisms, and of resistance in several host plants. However, there are many plant diseases which do not seem to fit the toxin model as we now understand it. We must remember that there are potential determinants of pathogenicity, virulence, and selectivity that are not toxic. For example, molecular determinants may be non-toxic but diffusible, or there may be chemical structures (toxic or non-toxic) associated with cell surfaces. The literature contains examples which seem to fit these categories (Oku et al., 1980; Sequeira, 1981).

II. Determinants of Plant Diseases

We must assume that disease development, host selectivity, and disease resistance have chemical bases. Where do the known toxins fit in this concept? Current understanding indicates that toxins are one of several types of disease determinants, which include pathogen-produced and host-produced factors. Known and hypothetical determinants produced by microorganisms include the following: (a) Low molecular weight toxins that interfere with metabolism or that change the structure of protoplasm. Some toxins are required for tissue colonization by certain pathogens. (b) Enzymes that break down cell walls or affect other cellular structures. (c) Hormone-like, or antihormonal compounds that may interfere with normal growth and development of the host plant. (d) Genetic information passed from pathogen to host. Once case is well-known, the T-DNA from *A. tumefaciens* (Kahl and Schell, 1982). (e) Non-toxic determinants involved in host-selectivity or in tissue colonization. This can include the pathogen's surface carbohydrates which may interact with specific proteins on the host, a phenomenon studied intensively for *Rhizobium*-legumes (Dazzo, 1980) and for *Pseudomonas solanacearum*-tobacco (Sequeira, 1981). Another possibility is the non-toxic, diffusable factor that is said to aid in tissue colonization and host selection by *Mycosphaerella pinodes* (Oku et al., 1980). (f) Products that may interfere with normal movement of water, nutrients, and metabolites.

The roles of T-DNA and certain toxins are well-defined, and the protein-carbohydrate relationship appears to be significant in several cases. Roles of the other pathogen-produced determinants appear not to be well-defined (Kosuge, 1981), although certain of them must have some part in disease development, at least in a secondary way. Host-produced compounds that may be involved in the interactions include phytoalexins, but again their roles are not precisely defined.

Several toxins are now known beyond reasonable doubt to be determinants in plant disease development. These toxins are often classified for convenience as either host-selective (or specific) and non-specific (Scheffer, 1976; Rudolph, 1976). The host-selective kinds are extremely toxic to hosts of the producing microorganism; non-host genotypes and species are tolerant. Several of the host-selective toxins are known to "recognize" plants which carry a single gene for sensitivity; this is the same gene that gives susceptibility to the producing fungus. Other toxins appear to "recognize" hosts at the species level, although the genetic control generally is unknown in these cases. Thus, there appear to be gradations in relative selectivity, which in some cases may match the selectivity of the producing microorganisms. The highly selective and the non-specific toxins are easily recognized as such, but there may be no clear or practical lines of distinction for intermediate types. Toxins that are thought to be non-specific sometimes show differences in relative toxicity to various plant species (Durbin & Uchytil, 1977; Owens & Wright, 1965), but this may not match the selectivity of the pathogen. Furthermore, some of the so-called non-specific toxins can be very selective for metabolic target sites (Steele et al., 1976), and this has added to the confusion in terminology.

It is sometimes useful to think of toxins as either "pathogenicity factors" or "virulence factors" (Yoder, 1980), using the meanings of these terms that are most familiar to plant pathologists (Scheffer & Briggs, 1981). Pathogenicity pertains to the ability of a microorganism to induce disease; a change in host range indicates a change in pathogenicity. Virulence refers to the relative severity of disease induced by a specific microorganism on a specified host. Genetic and other data indicate that most of the host-selective toxins are required for pathogenicity of the producing fungi (Scheffer, 1976); two well-known examples are the toxins from *H. victoriae* and *H. carbonum*. In contrast, genetic and mutant analyses indicate that certain non-specific toxins are not required for pathogenicity, but contribute to virulence and are responsible for certain symptoms; the best example is tabtoxin from *Pseudomonas syringae* pv. *tabaci* (Braun, 1937). In this case it is evident that factors other than tabtoxin also are significant in pathogenesis. This was evident from the genetic studies of Clayton (1947), who found that plants with monogenic resistance to *P. syringae* pv. *angulata* (non-producer of toxin) were also resistant to pv. *tabaci* (the toxin producer). Keep in mind that all these examples may be extreme cases; there may be other cases for which we have no clear lines of distinction between pathogenicity factors and virulence factors.

P. syringae pv. *phaseolicola* also produces a chlorosis-inducing toxin (phaseolotoxin), but this ability does not explain selectivity to bean. A polysaccharide produced by the bacterium could be the missing disease determinant. El-Banoby and Rudolph (1979) have reported that a specific polysaccharide from the bacterial slime will induce water congestion in bean leaves that are susceptible to the bacterium; the response in resistant bean leaves and in non-host species was minimal and quick to disappear. This water congestion is thought to aid the growth of bacteria in susceptible tissue. There was a correlation between the relative level of resistance to the pathogen and the relative sensitivity to the polysaccharide, and the polysaccharide was recovered from infected tissues. In time, the affected susceptible tissues apparently recover, making this an unusual toxin, if in fact it should be called a toxin. Confirmation of the work with additional bean cultivars, non-host species, and bacterial isolates is needed; nevertheless, the reports are potentially of great value for extending our concepts of chemical determinants of disease.

III. Known Toxins, Hypothetical Toxins and Toxin Artifacts

Relatively few toxins have received adequate evaluation for roles in disease development. The host-selective toxins known to date are listed in table 1; all of them are produced by fungi with specialized host ranges. Only a few of these toxins have been characterized chemically, but the existence of each toxin listed in table 1 has been demonstrated beyond reasonable doubt. In addition to the toxins which are selective at the genotype level, there are toxins which appear to have more limited specificity, expressed at the species level. For example, *Hypoxylon mammatum* produces toxic substances with high activity (in the ng range) and selectivity for the host species (*Populus tremuloides*); other plant species are very tolerant (Schipper, 1978; Stermer et al., 1981). However, there are tolerant and sensitive clones of *P. tremuloides*, and any possible correlation with disease resistance and susceptibility is unknown. More study will be needed to evaluate the role of *H. mammatum* toxins in disease.

The non-specific toxins known to be involved in disease are also few in number. Bacteria of the genus *Pseudomonas* are well-known for toxin production (table 2). These are

TABLE 1. Fungal-Produced Host-Selective Toxins Known to Date (1982). These toxins have the same genotype specificity as does the producing fungus.[a] Several have been characterized chemically, as indicated.* The cited references are not necessarily the original reports of the toxin, but are those that show a role in disease, or are review papers that bring information together.

Toxin[b]	Producing fungus	Plant affected	Reference[c]
1. HV	*Helminthosporium victoriae*	Oats	Scheffer 1976
2. HC*	*H. carbonum* race 1	Maize	Scheffer 1976
3. HS*	*H. sacchari*	Sugarcane	Steiner & Byther 1971
4. HmT*	*H. maydis* race T	Maize, Tms cytoplasm	Hooker et al. 1970, Yoder 1976
5. AK*	*Alternaria kikuchiana*[c]	Japanese pear	Nishimura et al. 1976
6. AM*	*A. mali*[c]	Apple	Kohmoto et al. 1977
7. AC-L	*A. citri*, lemon race[c]	Rough lemon	Kohmoto et al. 1979
8. AC-T	*A. citri*, tangerine race[c]	Tangerine, Mandarin	Kohmoto et al. 1979

9. AL*	*A. alternata f. lycopersici*	Tomato	Gilchrist & Grogan 1976
10. AF	*A. fragariae*[c]	Strawberry	Park et al. 1981
11. ALo	*A. alternata* f. *longipes*	Tobacco	Kohmoto et al. 1981
12. PC	*Periconia circinata*	Sorghum	Scheffer 1976
13. PM	*Phyllosticta maydis*	Maize, Tms cytoplasm	Yoder 1973, Comstock et al. 1973
14. CC	*Corynespora crassicola*	Tomato	Onesirosan et al. 1975

[a]In addition to these, the following fungi produce toxins which appear to be species-selective; (1) *Hypoxylon mammatum* affecting *Populus tremuloides* (Schipper, 1978); (2) *Alternaria eic-horniae* affecting water hyacinth (Maity & Samaddar, 1977). Certain bacterial polysaccharides from *Pseudomonas* and *Xanthomonas* spp. may be genotype-selective (See Table 2).

[b]Shorthand designations are followed (Scheffer & Briggs, 1981).

[c]All these *Alternaria* species are now considered to be form races of *Alternaria alternata*.

TABLE 2. Bacterial Toxins Involved in Disease Development. These toxins are chemically defined, except as indicated.

Toxin[a]	Producing bacterium	Plants infected	Reference
1. Tabtoxins	*Pseudomonas syringae*		
	pv. *tabaci*	Tobacco	Braun 1937
	pv. *coronofaciens*	Corn, oats, timothy	
	pv. *garcae*	Coffee	
2. Coronatine	*P. syringae* pv. *atropurpurea* and other forms	Grasses, soybean	Nishiyama et al. 1976
3. Phaseolotoxin & related compounds	*P. syringae* pv. *phaseolicola*	Bean	Mitchell & Bieleski 1977, Staskavicz & Panoupolus 1979
4. Rhizobitoxine	*Rhizobium japonicum*	Soybean	Owens & Wright 1965
5. Syringomycin[b]	*P. syringae* pv. *syringae*	Many species	DeVay et al. 1968, Sinden et al. 1971, Gonzalez & Vidaver 1979

6. Syringotoxin[b]	*P. syringae* pv. *syringae*	Citrus	Gonzalez et al. 1981
7. Tagetitoxin[b]	*P. syringae* pv. *tagetis*	Marigold	Mitchell & Durbin 1981
8. Polysaccharides[b,c]	*P. syringae* pv. *phaseolicola* & other forms	Bean	El Banoby & Rudolph 1979

[a]Several non-toxic compounds also are known determinants of disease, notably IAA from *P. syringae* pv. *savastanoi* (Smidt & Kosuge, 1978). Cytokinins from *Corynebacterium fasciens* may be involved in disease.

[b]Toxins not fully characterized, and roles not fully defined.

[c]These polysaccharides are reported to be host-selective.

TABLE 3. Fungal-Produced but Non-Selective Determinants of Plant Diseases. There are convincing evidences that these toxic compounds are involved in disease development. Chemical structures are known (Stoessl 1981).

Toxin	Producing fungus	Plant infected	Reference[a]
1. Fumaric acid	*Rhizopus* spp.	Almond	Mirocha et al. 1961
2. Fusicoccin	*Fusicoccum amygdali*	Almond	Ballio et al. 1976
3. Tentoxin	*Alternaria alternata* f. *tenuis*	Various seedlings	Templeton 1972
4. Oxalic acid	*Sclerotium rolfsii* *Sclerotinia sclerotiorum* & other spp.	Various plants	Noyes & Hancock 1981
5. Cyanide	Unidentified *Basidiomycete*	Alfalfa	Lebeau & Dickson 1955

[a]References are those which indicate a role in disease.

among our best-known toxins, especially for chemical characteristics and mechanisms of action. The non-specific toxins of fungal origin that are known conclusively as disease determinants are even fewer in number (table 3). These toxins vary from common organic acids to more complex molecules such as fusicoccin and tentoxin. Fusicoccin probably is the best-known and most studied of all toxins, and there has been some excellent work on action of tentoxin.

There are several characterized toxic metabolites, in addition to those listed in tables 1, 2, and 3, which appear to be involved in disease development (table 4). However, further evaluations are needed before firm conclusions can be drawn regarding the roles of compounds listed in table 4. Diverse types of compounds are represented in this group, as in the other groups; the most active of the compounds listed in table 4 are of considerable interest and potential for enlarging the toxin concept. Still another group of interest and potential are those compounds described as mycotoxins (table 5), but which also are toxic to plant tissues. Again, these compounds need careful evaluation for roles in plant disease development. Finally, there is a large group of compounds, of diverse types from various microorganisms, which are toxic to plants, but which have received no or very inadequate evaluations (table 6). Careful study could reveal roles for some of these compounds. Further information is available in Stoessl's review (1981).

The first major consideration of the researcher on toxins is to evaluate the role of each candidate toxin in disease development. Is the putative toxin required for infection to occur? Does it contribute to virulence of the producer? Does it account for any distinctive symptoms of the disease? Is it an artifact? Evaluations of biologically-active compounds have always been difficult; with toxins, this is perhaps the most difficult hurdle to overcome and is often overlooked. The literature contains many examples of meaningless work on toxins, and ill-founded claims continue to appear. In most cases, substances found in cultures are relativley low in toxicity, carry no distinctive markers, and have no distinctive effects. Generally, it has been impossible to relate such substances to the etiology of disease. Many lines of evidence have been used in toxin evaluations, which have been discussed elsewhere (Scheffer & Briggs, 1981) and which will be considered further in this volume. The use of mutants and genetic analyses of pathogens is especially important in toxin evaluation. If the objective is to understand disease development, then an adequate evaluation is essential.

TABLE 4. Fungal-Produced Compounds Which Appear to be Involved in Disease Development, But More Evaluations Are Needed.

Toxin	Producer	Plant infected	Reference[a]
1. Ascochyta toxin[b] (toxic at 0.2 μg/ml)	*Ascochyta chrysanthemi*	Chrysanthemum	Schadler & Bateman 1975
2. Cercosporin (a photosensitizing agent)	*Cercospora* spp.	Beet & other spp.	Calpouzos & Stalknecht 1967, Daub 1982
3. Dothiostromin (active at the 1 μg/ml level)	*Dothiostroma pini*	Pine	Shain & Franich 1981
4. Graminin A (limited host-selectivity & other suggestions of a role)	*Cephalosporium gramineum*	Wheat	Kobayashi & Ui 1979, Creatura et al. 1981
5. Glycopeptides[c]	Many microorganisms	Many species	Van Alfen & Allard-Turner 1979
6. Gregatins (limited host-selectivity)	*Cephalosporium gregatum*	Bean	Kobayashi & Ui 1977

7. Ophiobolins	*Helminthospprium oryzae* & other *H.* spp.	Rice, maize	Chattopadhyay & Samaddar 1976
8. Isomarticin & related compounds	*Fusarium martii* f. *pisi* & related fungi	Pea	Kern 1978
9. Gibberellins	*Gibberella zeae*	Rice	Pegg 1976
10. Polysaccharides[c]	Many microorganisms	Many plants	Van Alfen & McMillan 1982
11. Cerato-ulmin	*Ceratocystis ulmi*	Elm	Takai 1980

[a]References are those describing attempts to evaluate roles in disease.
[b]Chemical structure not determined.
[c]These compounds can cause plugging of xylem; i.e., their effects appear to be mechanical rather than biochemical.

TABLE 5. Examples of Mycotoxins (Toxic to Animals) That Are Also Toxic to Plants.[a] These compounds, produced by plant pathogens and saprophytes, have been found in various commodities, including fruit, grains, peanuts, and cottonseed, and in forage plants and plant debris. References are those that show toxicity to plants.

Toxic compounds	Producer	Reference
1. Aflatoxins	*Aspergillus flavus* & related fungi	Dashek & Llewellyn 1977
2. Citrinin	*Penicillium* spp., *Aspergillus* spp.	Damodaran et al. 1975
3. Moniliformin	*Fusarium moniliforme*	Cole et al. 1973
4. Oosporein	*Chaetomium trilaterale*	Cole et al. 1974
5. Patulin	*Penicillium, Aspergillus* & other spp.	Ellis & McCalla 1973
6. Penicillic acid	*Penicillium, Aspergillus* & other spp.	Sassa et al. 1971, Stoessl 1981
7. Rubratoxins & related compounds	*Penicillium, Byssochlymys* & other spp.	Reiss 1977
8. Sporidesmin	*Pithomyces chartarum*	Wright 1968
9. Trichothecenes[b]	*Fusarium* spp. & other fungi	Smalley & Strong 1974

[a] Zearalenone, an estrogenic compound produced by Fusarium spp., is found in moldy corn but is not known to be toxic to plants.

[b] Trichothecenes include diacetoxyscerpenol, diacetylnivalenol, epoxytrichothecenes, trichothecin, T-2 toxin, triacetoxyscerpenediol, trichodermin, and verrucarin.

TABLE 6. Toxic or Biologically Active Compounds Produced by Plant Pathogens, But With No Known Roles in Disease Development. Careful evaluation may show that some are disease determinants. These are examples only; many other compounds and microorganisms could be included in the list. For further information, see Stoessl (1981) and Rudolph (1976).

Toxic compound	Producer	Plant infected
1. Abscissic acid	*Cercospora rosicola*	Rose
2. Altenin	*Alternaria kikuchiana*	Japanese pear
3. Alternaric acid	*Alternaria solani*	Potato
4. Alternariols	*Alternaria* spp.	Various plants
5. Ascochitine	*Ascochyta* spp.	Pea
6. Colletotrichins	*Colletotrichum* spp.	Tobacco, etc.
7. Cytochalasins	*Phoma*, *Phomopsis*, etc.	Various plants
8. Cytokinins	*Corynebacterium fasciens*	Peas, etc.
9. Diaporthin	*Endothia parasitica*	Chestnut
10. Enniatins	*Fusarium* species	Various plants
11. Fomannosin	*Fomes annosus*	Pine
12. Fusaric acid	*Fusarium oxysporum*	Tomato, etc.
13. Lycomarasmin	*Fusarium oxysporum*	Tomato
14. Malformin	*Aspergillus niger*	Onion
15. Naphthazarins[a]	*Fusarium* spp.	Various plants
16. Phenylacetic acids	*Rhizoctonia solani*	Various seedlings
17. α-picolinic acid	*Pyricularia oryzae*	Rice
18. Piricularin, Pyriculol	*Pyricularia oryzae*	Rice
19. Rhynchosporosides	*Rhynchosporium secalis*	Barley
20. Sativanes[b]	*Helminthosporium* spp.	Wheat & other grains
21. Skyrin	*Penicillium* spp.	Various plants
22. Tenuazonic acid	*Alternaria* spp. *Pyricularia oryzae*	Various plants
23. Zinniol	*Alternaria zinniae*	Zinnia, marigold

[a]Fusarubin, javanacin, marticin, norjavanicin, novarubin.
[b]Helminthosporol, prehelminthosporal, victoxinine.

IV. Toxins as Factors in the Incidence of Disease

Toxins can be important factors in plant disease epidemics; the best case in point is the host-selective toxin from *Helminthosporium maydis* race T. The old race of the fungus (now known as race 0) produces no toxin known with assurance to be involved in disease, and was confined largely to southeastern U.S. and the southern fringes of the midwest corn belt. The reasons for this geographic limitation are not clear. The new race (race T, the toxin producer) appeared in the corn belt in 1968; race T is indistinguishable from race 0, except for the ability to produce the toxin that is specific to maize with Texas male sterile (Tms) cytoplasm. By 1970, *H. maydis* had spread throughout the corn belt, presumably as a result of its newly-acquired ability to produce a toxin that was destructive to the widely-planted Tms maize. *H. maydis* race T readily overwintered in the northern states, and attacked Tms maize again in 1971 (Ullstrup, 1972). When Tms maize was replaced by normal cytoplasm maize, the blight disease subsided and *H. maydis* disappeared in the northern zone of the corn belt. Another factor in this analysis is that race T apparently arose as a mutant from race 0, or was recently introduced into the area; the evidence is that the great epidemic of 1970 was incited by a fungus with only one of two possible mating types. In subsequent years, race T was blessed with both mating types (Leonard, 1973 & 1974). Many of the elements that resulted in the epidemic in maize probably were in effect for the continental epidemic of oats that was incited by *H. victoriae* in 1946-1948.

Epidemics caused by several producers of host-selective toxins were tied closely to the work of plant breeders. Their efforts sometimes culminate in introduction and widespread use of superior but genetically uniform cultivars, which may be vulnerable to new or highly adaptable pathogens. In a sense, the plant breeder is breeding new pests along with new crop genotypes. The first well-known example was the blight of oats caused by *H. victoriae* which followed widespread introduction and use of oats with the V_b gene for resistance to *Puccinia coronata*. The *H. maydis* (race T) blight of maize, summarized above, developed when most of the crop in the U.S. corn belt was converted to hybrids carrying Texas male sterile cytoplasm, used for economy in seed production. Epidemics caused by *Periconia circinata* in grain sorghum and *H. sacchari* in sugarcane, among others, have agronomic backgrounds comparable in some respects to that of the oat epidemic. These situations were discussed in more detail in an earlier review (Scheffer, 1976).

Diseases caused by fungi of the genus *Alternaria* are good examples showing that "new" plant diseases may arise when new, toxin-producing races appear, or when previously-localized, toxin-producing races become widespread. *A. citri* was known in Australia, Florida, and other areas as a benign pathogen on senile tissues of various citrus species. A virulent race of *A. citri* appeared in Australia some years ago (Pegg, 1966), and became locally destructive. This race, highly specialized to the Emperor mandarin, had the same morphology as the previously-known, non-specialized form of *A. citri;* the only obvious difference was the ability of the new race to produce a very toxic host-selective factor (Pegg, 1966; Kohmoto et al., 1979). The obvious hypothesis is that the new disease resulted from an ability, aquired by mutation or otherwise, to produce toxin. The same or a very similar toxin-producing race appeared in Florida in 1974 (Whiteside, 1976); the Florida race affects Dancy tangerine, which is closely related to the Emperor mandarin. We suspect that the Florida and Australia races had separate origins. A third race of *A. citri,* specalized to certain selections of rough lemon, was known (Whiteside, 1976); the lemon race also produces a toxin that is specific to its host (Kohmoto et al., 1979). Note that these new forms apparently appeared in populations of an existing although benign pathogen, not from saprophytes. The evolution of a pure saprophyte to a pathogen probably involves much more than simply the ability to produce a toxin; for example, saprophytes lack the ability to form penetration structures. Similar or even better examples of new toxin-producing forms of *Alternaria* have been studied in Japan, as will be discussed by Nishimura in a later chapter.

Non-specific toxins also can be important factors in epidemics, although the roles are less clear than with the host-selective toxins described above. The best example of a potential role of a non-specific toxin in epidemics is based on early work by Jensen & Livingston (1944) with bean blight caused by *Pseudomonas syringae* pv. *phaseolicola*. These researchers classified isolates of the bacterium by their ability to induce chlorotic halos; there were strong halo-inducers (later found to be toxin producers), weak or intermediate inducers, and non-inducers (non-producers of toxin). The isolates that caused pronounced chlorosis also caused severe stunting, wilt, vein clearing, systemic infections, and death of many plants. Isolates that caused lesions without yellow halos also gave fewer lesions, little or no stunting, no vein clearing, no wilting, and rarely killed plants. Members of the latter group were less effective than were the halo-inducers as pathogens in the field. A clue to the meaning of the work by Jensen and Livingston was evident when Patil et al. (1974) reported a pathogenic mutant which did not produce toxin, did not induce yellow halos, and did not invade plants systemically. The significance of toxin was clear when Staskawicz and Panopoulos (1979) reported a complete correlation between toxin-producing and chlorosis-inducing abilities of mutants of *P. syringae* pv. *phaseolicola*.

Two final examples will illustrate how knowledge of toxins can help in understanding seasonal occurrence of disease outbreaks. Temperatures above approximately 34°C cause sorghum and sugarcane tissues to become insensitive to toxins from *Periconia circinata* and *Helminthosporium sacchari*, respectively (Byther & Steiner, 1975; Bronson & Scheffer, 1977). Sensitivity returns in approximately three days at temperatures below 25°C. In the U.S. southwest, the sorghum disease was evident in the spring, disappeared during mid-summer, and reappeared in the fall (Quimby & Karper, 1949). Inoculated plants in growth chambers grew well with no symptoms at 35°C, but quickly blighted when temperatures were lowered to 22°C (Bronson & Scheffer, 1977). The sugarcane disease is evident in Hawaii and south Florida during the winter and early spring, but virtually disappears during the summer (Byther & Steiner, 1975; J.L. Dean, personal communication).

These examples illustrate the potential of toxin studies for understanding epidemic and seasonal occurrence of plant diseases. Clearly, a toxin can be the crucial factor that results in disease of epidemic proportions. Further knowledge of more toxins could lead to practical applications in plant pathology. So far, research along these lines has been extremely limited.

V. Toxicity and the Biochemistry of Toxic Action

What do toxins do to plants, and what are their mechanisms of action? This will be covered in detail in Chapter 3, but a quick introduction should be helpful to the reader's perspective. Our understanding at this time is fragmentary, although some excellent progress has been made on some of the non-specific toxins. Three known bacterial toxins (tabtoxin, phaseolotoxin, and rhizobitoxine) and a fungal toxin (tentoxin) cause striking chlorosis in green leaves of many test plant species. Each of these toxins is an effective enzyme inhibitor: tabtoxin inhibits glutamine synthetase (Uchytil & Durbin, 1980); phaseolotoxin inhibits ornithine carbamoyltransferase (Patil et al., 1970; Ferguson & Johnson, 1980); rhizobitoxine inhibits β-cystathionase (Giovanilli et al., 1973); and tentoxin inhibits coupling factor 1 in chloroplasts (Steele et al., 1976). However, it is not clear that these are the only sites of action *in vivo*, and it is not clear how these biochemical lesions lead to chlorosis in all cases. Nevertheless, these activities are among the best known of toxic effects. Much attention has been given to action of fusicoccin, which appears to affect cellular transport systems, notably the H^+/K^+ pumps (Marrè, 1977). Cercosporin is a photosensitizing agent which causes peroxidation of membrane lipids (Daub, 1982a & b).

The host-selective toxins cause all the visible and the known physiological effects that are induced in plants by the infecting fungi. These effects vary somewhat with the toxin, but all examined cases cause increases in respiration and ion leakage; other effects include decreases or increases in protein synthesis, uptake of selected solutes, and fixation of CO_2 in the dark. Most of the physiological changes are secondary to prime biochemical lesions, a conclusion based on genetic analyses and on experiments with isolated organelles (Scheffer, 1976). A hypothesis of long standing is that some of the toxins have receptor or sensitive sites in the susceptible cell, and that such sites are lacking in the resistant cell. The plasma membrane is most often suggested as the site of the receptor. This is still a viable hypothesis, but conclusive proof is lacking. There were reports of a receptor protein which binds toxin from *H. sacchari*, in the plasmalemma of sugarcane cells (Strobel, 1973). These claims are no longer tenable (Lesney et al., 1982). Mitochondria from susceptible maize are very sensitive to toxin from *H. maydis* race T, and there is evidence for a mitochondrial effect in intact tissues (Malone et al., 1978) and in protoplasts (Walton et al., 1979). Mitochondria from and in resistant maize tissues are not affected. The possibility of other prime biochemical lesions for HmT toxin has not been eliminated.

VI. Genetics and Toxicology

Genetic studies have been especially valuable for an understanding of plant toxins, as has been mentioned previously in this chapter. The genetic approach promises to be even more significant in the future. The power of the genetic tool is evident in some of the work on HC and HV toxins; key roles of these two toxins in disease were firmly established by genetic data on both the hosts and the pathogens (Scheffer, 1976).

A number of papers contain confirmatory data on the genetics of resistance and susceptibility of oats and maize to *H. victoriae* and *H. carbonum* race 1. All oat and maize genotypes that are susceptible to these fungi are sensitive to their respective toxins. All plant species and genotypes that are resistant to the fungi are tolerant of their toxins. Genotypes that are intermediate in resistance to the fungi are intermediate in senstivity to the toxins. Disease resistant and susceptible plants have been crossed; analyses of the progeny have shown that resistance and susceptibility in each case is controlled by one gene pair. Tolerance and sensitivity to each toxin is controlled by the same genes that control resistance and susceptiblity to the fungi. In oats, susceptiblity to *H. victoriae* and to its toxin is dominant over both resistance and intermediate. Thus, the range of reactions to the fungus and to its toxin is controlled by a gene locus (V_b) which appears to have two or more alleles with semi-dominance (Scheffer, 1976).

In maize, a major dominant gene (Hm) for resistance to *H. carbonum* race 1 is located on chromosome 1. The Hm gene has two known alleles (Hm^A and Hm^B) which give intermediate levels of resistance. A minor gene (Hm2) for resistance, located on chromosome 9, gives intermediate resistance, which again is dominant over susceptiblity (hm2). Different combinations of these genes and alleles result in plants that vary in reactions to the fungus, from very susceptible through several intermediate levels to very resistant (Nelson & Ullstrup, 1964). Three genotypes (hm/hm/-hm2/hm2; Hm^A/Hm^A-hm2/hm2; and Hm/-Hm2/-) were tested for sensitivity to HC toxin; they were susceptible, intermediate, and resistant, as predicted (Kuo et al. 1970). Other genotypes were not available for testing.

H. carbonum and *H. victoriae* were analyzed to determine the genetic control of pathogenicity and toxin production. All isolates which produced the proper toxin were pathogenic to the hosts that were susceptible to the producing fungus, without exceptions. All isolates that failed to produce such toxins in culture were non-pathogenic to both oats and maize. These relationships held with wild-type isolates and with mutants. Certain isolates of the two fungi are sexually compatible; the perfect stage belongs to the genus *Cochliobolus*. Matings of *C. victoriae* and *C. carbonum* race 1 gave progeny which produced either oat or maize-specific toxin, or both toxins, or neither toxin, in a 1:1:1:1 ratio. The progeny that produced only HV toxin were pathogenic only to oats with the V_b gene; those that produced only HC toxin were pathogenic only to maize of the appropriate genotypes; those that produced both toxins were pathogenic to both hosts; and those that produced neither toxin were non-pathogenic to oats and maize. Without exception, pathogenicity was correlated with toxin-producing ability, thus establishing key roles of these toxins in pathogenicity (Scheffer et al., 1967). An isolate that produces both toxins would appear to have increased potential for inciting epidemics; such isolates have never been found in nature.

Much confusion has resulted from considerations of toxins in relation to the gene-for-gene hypothesis which holds that certain genes for avirulence are specifically recognized by corresponding genes for resistance in the host. Such matched gene pairs always result in an incompatible (resistant) interaction regardless of other gene pairs that are present and that should specify compatibility (susceptibility) (Ellingboe, 1976). On the other hand, there are many plant diseases with only one known locus in control of resistance/susceptibility; resistance can be dominant or recessive. A complex gene-for-gene relationship has not been established in these cases. All known diseases involving host-selective toxins fit the simpler pattern, with resistance either dominant, semidominant, or recessive. The same host gene controls toxin sensitivity and disease reaction (Scheffer, 1976). There are no known examples of diseases involving toxins that fit a complex gene-for-gene relationship. The genetic patterns indicate that the known host-selective toxin cases probably could not be fit into the complex gene-for-gene pattern (Ellingboe, 1976), even if we assumed the production of slightly different toxins that affect different host genotypes. Thus, there may be important differences between the diseases that involve host-specific toxins and the diseases with a complex gene-for-gene pattern. Present knowledge indicates that toxin cases are as common as are complex gene-for-gene cases. However, the few in-depth studies preclude firm conclusions. Consider for example the case of the single gene that gives resistance to *Puccinia coronata* and susceptibility to *H. victoriae;* resistance/ susceptibility to *P. coronata* is thought to fit a complex gene-for-gene relationship.

VII. Wilt Toxins

A consideration of "wilt toxins" is important to our perspective, because the toxin theory was developed to account for symptoms of vascular wilt diseases (Hutchinson, 1913). Over the years, there were many reports of toxic culture filtrates from wilt pathogens; in hindsight, most of these reports need no further consideration. In some cases, host-selective activity was reported, but the claims have never been confirmed. The most intensive studies on wilt toxins, by Gäumann and associates (1954), resulted in discovery and characterization of several toxic compounds from Fusaria; best known are lycomarasmin and fusaric acid. However, even these findings soon became controversial. The general consensus today is that toxins are not the direct cause of wilting associated with xylem invasion by *Fusarium, Verticillium, Pseudomonas,* and *Ceratocystis* species, and others (Dimond, 1972). This conclusion is based on lack of definitive support for the wilt toxins, on accumulation of negative evidence (Kuo & Scheffer, 1964; Dimond, 1972), and on conclusive support for an alternative explanation.

A vast collection of data from many sources clearly shows that wilting in xylem-infected herbaceous and woody plants is caused by lack of water movement to leaves, related in turn to xylem plugging (Dimond, 1972; Duniway, 1973). The critical place for vascular dysfunction is the leaf petiole; for example, Duniway (1971) found that resistance to water flow in petioles gradually becomes almost total as wilt develops. Causes of blackage in different species may vary somewhat, but certainly include the pathogen itself (especially with bacteria), polysaccharides and other high mol wt excretions from the pathogen, host cell breakdown products, tyloses, and other host responses to the pathogens. Free polysaccharides appear to attach to vessel walls and pit membranes, thus reducing lateral movement of water to other vessels and to living cells (Dimond, 1972).

Refutation of toxins as the direct cause of wilt does not eliminate toxins as factors in systemic xylem diseases. Wilt is only one of several symptoms; observations and circumstantial evidence indicate the involvement of toxins in the overall syndromes. The vascular pathogens are confined until late in disease development to xylem tubes, which are the walls of dead cells. Other tissues, including leaf blades, are free of the invaders; nevertheless, there are drastic changes in the pathogen-free tissues. The simplest explanation is that the changes result from action of diffusable metabolites from the pathogen. Visible symptoms logically associated with toxic effects include chlorosis and stunting.

The following are examples of changes in plant cells that are not in direct contact with the wilt pathogens: a) There is a diseased-induced loss of electrolytes and an increase in respiration of leaves, soon after *Fusarium oxysporum* f. *lycopersici* invades the roots and stems of tomato plants (Collins & Scheffer, 1958). b) Tissues throughout the tomato plant become discolored well before severe symptoms show, and prior to contact with the pathogen. This early discoloration is evident in cell walls, but is seen only when tissues are cleared of chlorophyll (Scheffer & Walker, 1953). c) The metabolism of tobacco tissues, including leaves, is changed within 48 hours after *Pseudomonas solanacearum* is introduced into xylem tubes. Phenylalanine, tryptophan, and ammonia are increased several fold, and there is an increase in protein synthesis (Pegg & Sequeira, 1968). These kinds of changes are induced by toxins involved in other diseases. A systemic response is not apparent when strains of *P. solanacearum* specialized to other hosts are injected into xylem vessels. Experiments in which susceptible and resistant stocks and scions are grafted together also indicate that toxins are involved in the vascular wilt diseases (Keyworth, 1964).

One final point must be emphasized. The toxic compounds described to date from xylem-invading pathogens do not cause the physiological changes and typical symptoms associated with these wilt diseases (Kuo & Scheffer, 1964; Dimond, 1972). Toxins of significance in the vascular wilt dieases are not yet identified; they remain a challenge.

VIII. Toxin Producers: Parasites or Saprophytes?

Plant pathologists often have the notion that toxin-producing microorganisms are atypical pathogens and as such are not suitable models for study of plant infections (Brian, 1973). However, all available data indicate that this may not be true. Certainly nothing unusual is evident about the diseases caused by the producers of host-selective toxins, except that we know about the toxic metabolites. Genetic data for the pathogens and the hosts show the usual patterns for control of pathogenicity and disease resistance. The physiological changes in host tissues that are characteristic of plant diseases in general are induced by the toxins and by the producing fungi (Scheffer, 1976). Epidemiological patterns and disease cycles are not unusual. Overall, the indications are that the host-selective toxin producers are rather typical plant pathogens, and are no more saprophytic than are microorganisms involved in many other well-known plant diseases. The data for diseases involving non-specific toxins are more fragmentary, but again there are no solid indications that they are anything other than conventional plant pathogens. Of course, some of the non-specific toxins have restricted roles, such as the induction of special symptoms. The basis of the common notion about the unusual nature of diseases involving toxins is not at all clear.

Toxin-producing microorganisms are often referred to as "necrotrophs" (Scott, 1976). The concept is that such pathogens kill tissues in advance of colonization, and live as saprophytes on dead tissue. I can find no solid basis for this idea. Several diseases involving host-selective toxins have been examined in some detail by histological procedures. The fungal spores germinated, penetrated host surfaces, and colonized living tisues in a pattern typical of many obligate and non-obligate parasites. This was in clear contrast to the behavior of typical saprophytes on plant surfaces (Yoder & Scheffer, 1969; Comstock & Scheffer, 1973). In those cases tested, toxin is released by germinating spores or by hyphae soon after germination, but the amounts of toxin involved in these early stages must be extremely minute. There were no indications of "killing in advance" at the early stages of tissue colonization; in fact, at least one of the toxins at low concentrations is known to stimulate growth and, at both low and high concentrations, to increase the uptake of certain solutes (Yoder & Scheffer, 1973). Presumably, toxin concentrations must gradually increase in tissues at later stages of disease, until toxemia and "killing in advance" are evident.

The existence of "non-colonizing" (non-parasitic) but pathogenic microorganisms is often proposed. The most frequently-cited examples involve pathogens causing frenching disease of tobacco and yellow strapleaf disease of chrysanthemum (Woltz, 1978). Both were once thought to result from release of toxins or toxin precursors from soil microorganisms that do not invade tissues. However, yellow strapleaf is now associated with boron deficiency in chrysanthemum (Cox, 1974). The tobacco disease is similar to the chrysanthemum disease in many ways; a re-examination of the cause of frenching in tobacco is in order. Even *Periconia circinata* was once thought to be a non-colonizing pathogen in sorghum; this is now known to be untrue. Thus, it is doubtful that we have conclusive examples of non-parasitic but pathogenic microorganisms. Furthermore, the basis of the concept of necrotrophy in parasitism has not been clarified.

IX. Obstacles in Toxin Research

The major goals of plant toxicology have been to identify toxins, to isolate and characterize them chemically, to determine their modes of action, and to relate all this to our understanding of plant disease. What has inhibited or prevented us from reaching these goals? Not the least of the problems has been researchers who were overly eager to accept toxin possibilities, and others who were overly skeptical of the whole concept. Many reports of simplistic work with toxic culture filtrates have contributed to skepticism. Dogmatic attitudes, especially about seemingly competing concepts such as gene-for-gene systems (see Section VI) and phytoalexins, certainly have not been helpful in understanding pathogen-produced determinants of disease.

A history of "wilt toxins" (see Section VII) will illustrate the problem with over-extension of the toxin concept. The wilt toxins were the basis of much biological and chemical work, and the controversies that developed led to a general skepticism. A weakness of the wilt toxin work appears to have been too much reliance on shoot cutting assays for toxicity. Many substances that accumulate in culture fluids will damage cut shoots from plants because all solutes are indiscriminately drawn in by transpirational pull. Transpiration also concentrates the solutes in leaves, leading to damage by compounds with low inherent toxicity, and to masking of effects of more interest. Polysaccharides often accumulate in cultures, and any large polysaccharide molecule can interfere with water movement in cuttings (Van Alfen and McMillan, 1982); this may or may not have parallels to what happens in the infected plant. Interference with water supply causes leaves to wilt, but this probably has nothing to do with ability of the pathogen to colonize tissues in the first place. Deficiencies in the wilt toxin idea probably contributed to skepticism for the whole toxin concept. Skeptical attitudes are reflected in statements such as those concerning the "off-beat" nature of host-selective toxins (Brian, 1973). Problems such as those encountered in the wilt toxin work further emphasize the importance of careful evaluation of each putative toxin.

Other obstacles have been of more practical nature. Lack of sensitive and reliable assays for toxins has on occasion been a major problem which made the difference between success and failure; development of a good assay is worthy of much effort. Chemical assays are desirable but bioassays are needed until chemical assays are available, usually only after the toxin is characterized. The best qualitative assays have been based on distinctive effects such as host selectivity. The best quantitative assays also rely on distinctive effects: for example, the inhibition of CO_2 fixation in the dark by HmT toxin. Less distinctive effects (such as toxin-induced chlorosis, electrolyte losses, and root growth inhibition) can be used if suitable precautions are taken. Use of proper controls cannot be overemphasized. For host-selective toxins, controls consisting of resistant tissues should always be used, and use of inactivated toxin is desirable. With proper controls, reliable results can be obtained even with impure toxin preparations, as has been illustrated many times. The object, of course, is to obtain the highly purified toxin preparations required for characterization and action studies.

Purification of toxins can be a difficult and time-consuming process. The researcher must always keep in mind that the critical toxins may be very active but labile, that they may be bound and inactivated in plant tissues, or that the effects of the crucial compounds may be masked by many other metabolic products of the pathogen. Modern and gentle methods of separation, such as chromatography of various types (TLC, HPLC, gel filtration, flash chromatography, etc.) have been great assets in purification of natural products. A reseracher with a good assay and with experience in the art and science of purification should no longer encounter great problems in purification of toxins.

Modern methods of chemical characterization have stimulated increased interest in toxins and other natural products. Widespread availability of MS, ^{1}H-NMR, ^{13}C-NMR, and other analytical instruments has greatly reduced the time required for characterization, and enlarged the possibilities for understanding toxin structures. In the future, we can expect routine characterization of toxins when they are isolated to a reasonably pure state and when their significance in disease is established.

New obstacles to research will no doubt develop, but we are gradually overcoming those that have been obvious in the past. Solid new data and concepts are appearing at increasing rates, as indicated by the dates of major accomplishments in the chronology of plant toxicology (Section X).

X. A Chronology of Plant Toxicology

It is customary to credit Anton de Bary with the idea that plant disease develoment is based on substances released from the pathogen. Clearly, the credit is justified as far as enzymes are concerned. Other names associated with the formative period of disease enzymology and plant toxicology include H. M. Ward of Britain and L. R. Jones of Wisconsin. In the ensuing years, there were many unsuccessful attempts to relate disease development with pathogen-produced factors other than enzymes. In general, this was a period of frustration, but some of the early contributions were of value as building blocks for later work (see Scheffer & Briggs, 1981). Eventually, a period of slow progress was underway.

I have prepared a three-part chronology of research contributions which in my judgment were significant in development of plant toxicology. The first part of the list includes initial reports of significant toxins plus descriptions of work aimed at evaluation of roles of specific toxins in disease development. All of plant toxicology is based on such pioneering efforts. Next, I have included a list of contributions which are significant for understanding the chemical structures of the toxins known to have roles in disease development. Some of the proposed structures have been amended by later work, and this is indicated. The third category in the list covers significant contributions aimed at understanding toxic action. Obviously there are subjective elements in choices of papers cited.

A. *Pioneering Contributions and Evaluations of Toxins*

1) Johnson & Murwin (1925) reported the existence of a chlorosis-inducing toxin from *Pseudomonas syringae* pv. *tabaci*, a pathogen of tobacco. The report was confirmed and extended by Clayton (1934), but there were only preliminary attempts at isolation. This was the first solid evidence that a pathogen-produced substance can be a determinant of a plant disease. The toxin is now known as tabtoxin.

2) Tanaka (1933) reported host-selective toxic activity in culture fluids of *Alternaria kikuchiana*, a pathogen of Japanese pears. There had been previous reports of selective toxicity, but none prior to Tanaka's was confirmed.

3) Braun (1937) found that isolates of *P. syringae* pv. *tabaci* which lost tabtoxin-producing ability could still be pathogenic, that the mutants were identical to *P. syringae* pv. *angulata*, and that several pathogenic but toxin-less mutants caused no chlorotic halos. These results indicated that the toxin is not required for infectivity or pathogenicity, but contributes to virulence and is responsible for the special symptoms. Braun's conclusions were supported by the work of Clayton (1947), who found that a gene for resistance to *P. syringae* pv. *tabaci* also gave resistance to pv. *angulata*. An analysis of mutants, such as that supplied by Staskawicz and Panopoulos (1979) for pv. *phaseolicola*, would add the final touch.

4) Meehan & Murphy (1947) reported selective toxicity by culture fluids of *Helminthosporium victoriae* to genotypes of oats that are susceptible to the fungus. The gene which gave susceptibility to *H. victoriae* also gave resistance to *Puccinia coronata*. The work was soon confirmed and was extended by Luke & Wheeler (1955).

5) Tabtoxin was isolated from cultures of *P. syringae* pv. *tabaci* by Woolley et al. (1952). This was the first authentic toxin to be purified and characterized (in part). However, the suggested structure did not prove to be correct and was later withdrawn.

6) *P. syringae* pv. *phaseolicola* a pathogen and chlorosis-inducer in bean, was reported by Waitz and Schwartz (1956) to produce a chlorosis-inducing toxin. The work was confirmed and extended by Hoitink et al. (1966). This toxin later became known as phaseolotoxin.

7) The toxin of *H. victoriae* was isolated to a relatively pure state and was partially characterized by Pringle & Braun (1957, 1958). This was the first serious attempt at isolation of a host-selective toxin. There were insufficient data to suggest a structure.

8) A host-selective toxin from *Periconia circinata*, cause of a blight in certain cultivars of grain sorghum, was reported and evaluated by Scheffer & Pringle (1961). Pringle & Scheffer (1966) later crystallized the toxin and reported some of its characteristics. Dunkle and associates have continued the work (Wolpert & Dunkle, 1980).

9) Mirocha et al. (1961) described exemplary work leading to a firm conclusion that fumaric acid is a determinant of an almond disease incited by *Rhizopus* spp.

10) Fusicoccin was reported as a product of *Fusicoccum amygdali*, another pathogen of almond, by Graniti (1964). The compound was isolated by Ballio et al. (1964, 1968). Action of the toxin has been the subject of intensive work, of interest to many plant physiologists. However, no serious attempts to evaluate the role of fusicoccin in disease were reported until 1976, when Ballio et al. found high concentrations of the very toxic compound in diseased tissues.

11) A host-selective toxin from *H. carbonum* race 1, a maize pathogen, was reported by Scheffer and Ullstrup (1965). The toxin was isolated and some of its characteristics were determined by Pringle & Scheffer (1967). Matings between *Cochliobolus* (*Helminthosporium*) *victoriae* (producer of oat-selective toxin) and *C. carbonum* (producer of maize-selective toxin) gave progeny which produced oat (HV) toxin, maize (HC) toxin, both toxins, or neither toxin, in a 1:1:1:1 ratio (Scheffer et al., 1967). Pathogenicities in all cases were correlated with toxin-producing abilities. These and other data led to the firm conclusion that the toxins are required for pathogenicity by the producing fungi, and that single genes can control toxin production. HC toxin is an example of a chemically defined final product (Liesch et al., 1982) of a gene for pathogenicity.

12) Owens & Wright (1965) reported that certain isolates of *Rhizobium japonicum*, a symbiont of soybeans, will cause yellowing of leaves. The same isolates were found to produce a chlorosis-inducing toxin (rhizobitoxine). The toxin was isolated and later characterized.

13) *A. alternata* f. *tenuis*, which causes chlorosis in seedling plants of many species, was shown by Fulton et al. (1965) to produce a chlorosis-inducing substance which later became known as tentoxin.

14) *A. mali*, a pathogen on certain cultivars of apple in Japan, was found by Sawamura (1966) to produce a toxin that selectively damages tissues of host cultivars. This toxin is now known as AM toxin.

15) DeVay et al. (1968) reported that *Pseudomonas syringae* pv. *syringae* produces syringomycin. The case for a role of this toxin in disease was presented here and in other papers from DeVay's laboratory.

16) HV toxin from *H. victoriae* was shown by Yoder & Scheffer (1969) to be involved in initial colonization of host tissue by the fungus. The toxin also was reported to cause the early physiological changes characteristic of infection. The behavior of *H. victoriae* differed significantly from that of saprophytes on plant surfaces.

17) Culture fluids of *H. maydis* race T were selectively toxic to maize with Tms cytoplasm, as reported first by Hooker et al. (1970). This original, preliminary report was quickly confirmed and the work was extended in several other laboratories.

18) Steiner & Byther (1971) reported that *H. sacchari* produces a toxin that affects susceptible but not resistant sugarcane.

19) Studies by Comstock & Scheffer (1973) supported earlier indications that HC toxin is required for colonization of susceptible maize by *H. carbonum*. The work also showed that dead or seriously damaged cells are not required for colonization (in contrast to the situation with true saprophytes), and that inhibitory substances from host cells do not account for resistance of maize to *H. carbonum* or to related pathogens.

20) A host-selective toxin from *A. alternata* f. *lycopersici*, a pathogen of tomato in California, was reported by Gilchrist & Grogan (1976). The same fungus was later found in Japan (Kohmoto et al., 1982).

21) A race of *A. citri* which is pathogenic to rough lemon in Florida was shown by Kohmoto et al. (1979) to produce a host-selective toxin. Another race of *A. citri*, pathogenic to Dancy tangerine, was also shown to produce a host-selective toxin. The toxins were isolated but not completely characterized. A toxin comparable to the one affecting Dancy tangerine was reported earlier by Pegg (1966) to be specific to Emperor mandarin.

22) Staskawicz & Panopoulos (1979), working with many mutants of *P. syringae* pv. *phaseolicola*, found a complete correlation between ability to produce phaseolotoxin and ability to cause chlorosis in bean. The data establish a firm role for phaseolotoxin in disease development.

23) Toxin production by *H. maydis* race T is controlled by a single gene, and the final product (HmT toxin) has been defined chemically (Tegtmeier et al., 1982; Yoder & Gracen, 1975).

B. *Contributions to an Understanding of Chemical Structures of Toxins*

1) Woolley et al. (1952) published the first isolation and characterization of a pathogen-produced toxic substance that clearly is a disease determinant in plants. The suggested structure of tabtoxin from *P. syringae* pv. *tabaci* had some problems and was later withdrawn.

2) The first attempt to characterize a host-selective toxin, that of *H. victoriae*, was described by Pringle & Braun (1957, 1958). The characterization was not sufficient to suggest a complete structure. Victoxinine, a sesquiterpene thought to be part of the toxin molecule, was characterized by Dorn and Arigoni (1972).

3) Ballio et al. (1968) published a structure of fusicoccin, a complex terpenoid from *Fusicoccum amygdali*.

4) The structure of tabtoxin, a β-lactam containing threonine and a novel amino acid (tabtoxinine), was determined by Stewart (1971).

5) A structure for rhizobitoxine from *Rhizobium japonicum* was submitted by Owens et al. (1972). The toxin, a chlorosis-inducer, is an enol-ether amino acid.

6) The host-selective toxin from *A. mali* was characterized and a structure was submitted by Okuno et al. (1974). This was the first complete characterization of a host-selective toxin. The structure, a cyclic depsipeptide, was confirmed by Ueno et al. (1975) and was synthesized by Lee et al. (1976).

7) Meyer et al. (1975) determined the structure of tentoxin (a cyclic tetrapeptide), largely by ^{1}H-NMR spectroscopy.

8) Mitchell (1976) characterized phaseolotoxin as (N^2 phosphosulphamyl)ornithylalanylhomoarginine.

9) Kono and Daly (1979) elucidated the structure of HmT toxin, a linear polyketol from *H. maydis* race T. Compounds with the same toxicity were synthesized by Suzuki et al. (1982).

10) The host-selective toxin of *A. alternata* f. *lycopersici*, which affects certain cultivars of tomato, was characterized as a 1-aminodimethylheptadecapentol (Bottini & Gilchrist, 1981; Bottini et al., 1981).

11) Host-selective toxin from *H. sacchari* was characterized in large part by Livingston & Scheffer (1981). Macko et al. (1981) presented further data and suggested a structure. Toxin is a β-1,4 galactofuranoside with four galactose units and a sesquiterpene. This toxin had previously been described as helminthosporoside, a cyclopropyl-α-pyranoside (Steiner & Strobel, 1971).

12) Liesch et al. (1982) suggested a structure for HC toxin from *H. carbonum* race 1. Toxin is a cyclic tetrapeptide with an epoxide group. Walton et al. (1982) and Gross et al. (1982) confirmed most of the structure, but suggested a different sequence of amino acids in the molecule.

13) Host-selective toxins from *A. kikuchiana* were characterized by Nakashima et al. (1982).

C. *Contributions to an Understanding of Toxic Action*

1) Braun (1950) described a pioneering effort to determine the action of the first known determinant of a plant disease. He tentatively concluded that tabtoxin is an antimetabolite of methionine; later work has not confirmed this.

2) Selective toxin from *H. victoriae* quickly induces losses of electrolytes from oat tissue (Wheeler & Black, 1963); this indicates drastic cellular damage. Samaddar & Scheffer showed that the effect was on the plasmalemma of oat cells with the V_b gene. It was not determined whether the effect was direct or secondary.

3) Tabtoxin was reported by Sinden & Durbin (1968) to inhibit glutamine synthetase, but this report was in doubt for many years. Later, tabtoxin was shown to be converted to tabtoxin-β-lactam, which is the enzyme inhibitor (Uchytil & Durbin, 1980; Turner, 1981).

4) Miller & Koeppe (1971) found that HmT toxin from *H. maydis* race T has a striking effect on mitochondria isolated from susceptible maize seedlings; mitochondria from resistant plants were not affected. The report was soon confirmed; later work indicated that toxin acts as an unusual uncoupling agent.

5) Fusicoccin was found by Turner (1973) to affect H^+/K^+ pumps in cells. The work was confirmed and extended by Marrè (1977, 1979) and others.

6) Giovanilli et al. (1973) presented evidence that rhizobitoxine may act by inhibiting β-cystathionase *in vivo*. Earlier work (Owens et al., 1968) had indicated that the enzyme from non-toxin-producing bacteria is inhibited *in vitro*.

7) HC toxin from *H. carbonum* increases the uptake of nitrate and certain other solutes by maize tissue, as reported by Yoder & Scheffer (1973).

8) Steele et al. (1976) found that tentoxin has a direct effect on chloroplast coupling factor 1; i.e., the toxin is an enzyme inhibitor.

9) Phaseolotoxin was shown by Mitchell & Bieleski (1977) to be converted by plant cells to phosphosulfamylornithine, which is the active toxin. Earlier work had shown that phaseolotoxin causes accumulation of ornithine in tissues, as occurs in infected plants (Patel & Walker, 1963). Patil et al. (1970, 1972) had indicated that the toxin inhibits ornithine carbomoyltransferase. Ferguson and Johnson (1980) extended and clarified the work, and confirmed that purified toxin causes ornithine to accumulate in tissues and inhibits

ornithine carbomoyltransferase. The enzyme from toxin-producing bacteria is resistant to the toxin (Ferguson et al., 1980; Staskawicz et al., 1980).

10) *H. maydis* race T toxin apparently acts on mitochondria in intact protoplasts, resulting in a decrease in ATP levels, according to Walton et al. (1979). Malone et al. (1978) had given earlier evidence that HmT toxin affects mitochondria in intact tissues. Electron microscopy confirmed an early, selective effect on mitochondria in tissues and protoplasts (Aldrich et al., 1977; Gregory et al., 1980). Other possible sites for direct action were not excluded.

11) Fusicoccin appears to bind in a meaningful way to receptor sites in oat cell membranes, according to Stout & Cleland (1980).

12) Tabtoxin as such is not active, but is hydrolyzed in plant tissues to tabtoxin-β-lactam, which inhibits glutamine synthetase (Uchytil & Durbin, 1980). Chlorosis of tissues is thought to result from release of ammonia in the reaction, according to Turner (1981), Turner & Debbage (1982), and Frantz et al. (1982).

XI. Concluding Remarks

Many toxic compounds are produced by microorganisms in culture, but only a few are established as determinants of disease (i.e., toxins). Some known toxins, which may be either host-selective or non-specific as to species affected, are major factors in development of destructive diseases of plants. Most of the host-selective toxins are considered to be "pathogenicity factors"; i.e., they are required by the producing microorganism to colonize tissue and induce disease. Some non-specific toxins may be required for pathogenicity; generally, however, the ones known to date are thought to contribute to virulence without being required for pathogenicity.

At least four categories of research on toxins are seen as important for progress in the near future. (a) There must be more concern with evaluation of each potential toxin for its possible role in disease. Genetic analysis has been a powerful tool for this purpose; it is expected to be still more useful in the future. (b) Knowledge of the chemical characteristics and structures of more toxins is needed. This should soon be forthcoming, because of the widespread availability of modern methods of analysis. (c) Work toward an understanding of mechanisms of action by each toxin is essential. This has been a difficult and frustrating area of research, but several notable successes suggest that the problems may now be more tractable. (d) Finally, we need to elucidate the roles of toxins in epidemics and in the seasonal incidence of plant diseases. To date, work on ecological and epidemiological aspects of plant toxicology has been extremely limited.

References

Aldrich, H.C., Gracen, V.E., York, D., Earle, E.D., and Yoder, O.C. (1977). *Tissue and Cell* 9, 167.

Ballio, A. et al. (1964). *Nature* 203, 297.

Ballio, A. et al. (1968). *Experientia* 24, 631.

Ballio, A. et al. (1976). *Physiol. Plant Pathol.* 8, 163.

Bottini, A.T., and Gilchrist, D.G. (1981). *Tetrahedron Letters* 22, 2719.

Bottini, A.T., Bowen, J.R., and Gilchrist, D.G. (1981). *Tetrahedron Letters* 22, 2723.

Braun, A.C. (1937). *Phytopathology* 27, 283.

Braun, A.C. (1950). *Proc. Nat. Acad. Sci. (U.S.)* 36, 423.

Brian, P.W. (1973). *In* "Fungal Pathogenicity and the Plant's Response" (R.J.W. Byrd and C.V. Cutting, eds.), p. 469. Academic Press, New York.

Bronson, C.R., and Scheffer, R.P. (1977). *Phytopathology* 67, 1232.

Byther, R.S., and Steiner, G.W. (1975). *Plant Physiol.* 56, 415.

Calpouzos, L., and Stallknecht, G.F. (1967). *Phytopathology* 57, 799.

Chattopadhyay, A.K., and Samaddar, K.R. (1980). *Phytopath. Z.* 98, 118.

Clayton, E.E. (1934). *J. Agr. Res.* 48, 411.

Clayton, E.E. (1947). *J. Heredity* 38, 35.

Cole, R.J. et al. (1973). *Science* 179, 1324.

Cole, R.J. et al. (1974). *J. Agr. Food Chem.* 22, 517.

Collins, R.P., and Scheffer, R.P. (1958). *Phytopathology* 48, 349.

Comstock, J.C., and Scheffer, R.P. (1973). *Phytopathology* 63, 24.

Comstock, J.C., Martinson, C.A., and Gengenbach, B.G. (1973). *Phytopathology* 63, 1357.

Cox, R.S. (1974). *Proc. Florida State Hort. Soc.* 87, 438.

Creatura, P.J., Safir, G.R., Scheffer, R.P., and Sharkey, T.D. (1981). *Physiol. Plant Path.* 19, 313.
Damodaran, C. et al. (1975). *Experientia* 31, 1415.
Dashek, W.V., and Llewellyn, G.C. (1977). *Ann. Nutr. Aliment.* 31, 841.
Daub, M.E. (1982a). *Phytopathology* 72, 370.
Daub, M.E. (1982b). *Plant Physiol.* 69, 1361.
Dazzo, F.B. (1980). *In* "The Cell Surface: Mediator of Developmental Processes" (S. Subtelny and W. Wessells, eds.), p. 277. Academic Press, New York.
DeVay, J.E. et al. (1968). *Phytopathology* 58, 95.
Dimond, A.E. (1972). *In* "Phytotoxin in Plant Diseases" (R.K.S. Wood, A. Ballio, and A. Graniti, eds.), p. 289. Academic Press, London.
Dorn, F., and Arigoni, D. (1972). *J. Chem. Soc. Chem. Comm.* 1972, 1342.
Duniway, J.M. (1971). *Nature* 230, 252.
Duniway, J.M. (1973). *Phytopathology* 63, 458.
Durbin, R.D., and Uchytil, T.F. (1977). *Phytopathology* 67, 602.
El-Banoby, F.E., and Rudolph, K. (1979). *Phytopath. Z.* 95, 38.
Ellingboe, A.H. (1976). *In* "Physiological Plant Pathology" (H. Heitefuss and P.H. Williams, eds.), p. 761. Springer-Verlag, Berlin and New York.
Ellis, J.R., and McCalla, T.M. (1973). *Appl. Microbiol.* 25, 562.
Ferguson, A.R., and Johnston, J.S. (1980). *Physiol. Plant Path.* 16, 269.
Ferguson, A.R., Johnston, J.S., and Mitchell, R.E. (1980). *FEMS Microbiol. Lett.* 7, 123.
Frantz, T.A., Peterson, D.M., and Durbin, R.D. (1982). *Plant Physiol.* 69, 345.
Fulton, N.D., Bollenbacker, K., and Templeton, G.E. (1965). *Phytopathology* 55, 49.
Gäumann, E. (1954). *Endeavour* 13, 198.
Gilchrist, D.G., and Grogan, R.G. (1976). *Phytopathology* 66, 165.
Giovanilli, J., Owens, L.D., and Mudd, S.H. (1973). *Plant Physiol.* 51, 492.
Gonzalez, C.F., and Vidaver, A.K. (1979). *Current Microbiol.* 2, 75.
Gonzalez, C.F., DeVay, J.E., and Wakeman, R.J. (1981). *Physiol. Plant Path.* 18, 41.
Graniti, A. (1964). *Phytopath. Medit.* 3, 125.
Gregory, P., Earle, E.D., and Gracen, V.E. (1980). *Plant Physiol.* 66, 477.

Gross, M.L. et al. (1982). *Tetrahedron Letters* 23, 5381.
Harper, J.R., and Balke, N.E. (1981). *Plant Physiol.* 68, 1349.
Hoitink, H.A.J., Pelletier, R.L., and Coulson, J.G. (1966). *Phytopathology* 56, 1062.
Hooker, A.L., Smith, D.R., Lim, S.M., and Beckett, J.B. (1970). *Plant Dis. Rep.* 54, 708.
Hutchinson, C.M. (1913). *Mem. Dept. Agric., India, Bact. Ser.* 1, 67.
Jensen, J.H., and Livingston, J.E. (1944). *Phytopathology* 34, 471.
Johnson, J., and Murwin, H.F. (1925). *Wis. Agric. Exp. Sta. Res. Bull.* 62, 1-35.
Kahl, G., and Schell, J.S. (eds.) (1982). "Molecular Biology of Plant Tumors." Academic Press, New York.
Kern, H. (1978). *Ann. Phytopathol.* 10, 327.
Keyworth, W.G. (1964). *Ann. Appl. Biol.* 54, 99.
Kobayashi, K., and Ui, T. (1977). *Physiol. Plant Path.* 11, 55.
Kobayashi, K., and Ui, T. (1979). *Physiol. Plant Path.* 14, 129.
Kohmoto, K., Taniguchi, T., and Nishimura, S. (1977). *Ann. Phytopath. Soc. Japan* 43, 65.
Kohmoto, K., Scheffer, R.P., and Whiteside, J.O. (1979). *Phytopathology* 69, 667.
Kohmoto, K., Hoshotani, Y., Otani, H., and Nishimura, S. (1981). *Proc. Phytopath. Soc. Japan, Annual Meeting,* Abstract no. 2-34.
Kohmoto, K. et al. (1982). *J. Fac. Agric., Tottori Univ.* 27, 1.
Kono, Y., and Daly, J.M. (1979). *Biorganic Chemistry* 8, 391.
Kosuge, T. (1981). *In* "Carbohydrates in Plant-Pathogen Interactions" (W. Tanner and F.A. Loewus, eds.), p. 584. Springer-Verlag Co., Berlin and New York.
Kuo, M.-S., and Scheffer, R.P. (1964). *Phytopathology* 54, 1041.
Kuo, M.-S., Yoder, O.C., and Scheffer, R.P. (1970). *Phytopathology* 60, 365.
Lebeau, J.B., and Dickson, J.G. (1955). *Phytopathology* 45, 667.
Lee, S., Aoyagi, H., Shimohigashi, Y., and Izumiya, N. (1976). *Tetrahedron Letters* 1976, 843.
Leonard, K.J. (1973). *Phytopathology* 63, 112.
Leonard, K.J. (1974). *Plant Dis. Rep.* 58, 529.
Lesney, M.S., Livingston, R.S., and Scheffer, R.P. (1982). *Phytopathology* 72, 844.

Liesch, J.M., Sweeley, C.C., Staffeld, G.D., Anderson, M.S., Weber, D.J., and Scheffer, R.P. (1982). *Tetrahedron* 38, 45.
Livingston, R.S., and Scheffer, R.P. (1981). *J. Biol. Chem.* 256, 1705.
Luke, H.H., and Wheeler, H.E. (1955). *Phytopathology* 45, 453.
Macko, V., Goodfriend, K., Wachs, T., Renwick, J.A.A., Acklin, W., and Arigoni, D. (1981). *Experientia* 37, 923.
Maity, B.R., and Samaddar, K.R. (1977). *Phytopath. Z.* 88, 78.
Malone, C.P., Miller, R.J., and Koeppe, D.E. (1978). *Physiol. Plantarum* 44, 21.
Marrè, E. (1977). *In* "Regulation of Cell Membrane Activities in Plants" (E. Marrè and O. Ciferri, eds.), p. 185. Elsevier Publ. Co., Amsterdam.
Marrè, E. (1979). *Ann. Rev. Plant Physiol.* 30, 273.
Meehan, F.L. (1951). *Iowa State J. Sci.* 25, 292.
Meehan, F., and Murphy, H.C. (1947). *Science* 106, 270.
Meyer, W.L. et al. (1975). *J. Amer. Chem. Soc.* 97, 3802.
Miller, R.J., and Koeppe, D.E. (1971). *Science* 173, 67.
Mirocha, C.J., DeVay, J.E., and Wilson, E.E. (1961). *Phytopathology* 51, 851.
Mitchell, R.E. (1976). *Phytochemistry* 15, 1941.
Mitchell, R.E., and Bieleski, R.L. (1977). *Plant Physiol.* 60, 723.
Mitchell, R.E., and Durbin, R.D. (1981). *Physiol. Plant Path.* 18, 157.
Nakashima, T., Ueno, T., and Fukami, H. (1982). *Tetrahedron Letters* 1982, 4469.
Nelson, O.E., and Ullstrup, A.J. (1964). *J. Heredity* 55, 195.
Nishimura, S., Kohmoto, K., Otani, H., Fukami, E., and Ueno, T. (1976). *In* "Biochemistry and Cytology of Plant-Parasite Interaction" (K. Tomiyama et al., eds.), p. 94. Elsevier Publ. Co., Amsterdam.
Nishiyama, K. et al. (1976). *Ann. Phytopath. Soc. Japan* 42, 613.
Noyes, R.D., and Hancock, J.G. (1981). *Physiol. Plant Path.* 18, 123.
Oku, H., Shiraishi, T., Ouchi, S., and Ishiura, M. (1980). *Naturwissenschaften* 67, 310.
Okuno, T., Ishita, Y., Sawai, K., and Matsumoto, T. (1974). *Chem. Letters (Chem. Soc. Japan)* 1974, 635.
Onesirosan, P. et al. (1975). *Physiol. Plant Path.* 5, 289.
Owens, L.D., and Wright, D.A. (1965). *Plant Physiol.* 40, 927.

Owens, L.D., Guggenheim, S., and Hilton, J.L. (1968). *Biochim. Biophys. Acta* 158, 219.
Owens, L.D., Thompson, J.F., Pitcher, R.G., and Williams, T. (1972). *J. Chem. Soc. Chem. Comm.* 1972, 714.
Park, P., Nishimura, S., Kohmoto, K., and Otani, H. (1981). *Ann. Phytopath. Soc. Japan* 47, 488.
Patel, P.N., and Walker, J.C. (1963). *Phytopathology* 53, 522.
Patil, S.S., Tam, L.Q., and Sakai, W.S. (1972). *Plant Physiol.* 49, 803.
Patil, S.S., Hayward, A.C., and Emmons, R. (1974). *Phytopathology* 64, 590.
Pegg, G.F. (1976). *In* "Physiological Plant Pathology" (R. Heitefuss and P.H. Williams, eds.), p. 592. Springer-Verlag, Berlin and New York.
Pegg, G.F., and Sequeira, L. (1968). *Phytopathology* 58, 476.
Pegg, K.G. (1966). *Queensland J. Agr. Anim. Sci.* 23, 15.
Pringle, R.B., and Braun, A.C. (1957). *Phytopathology* 47, 369.
Pringle, R.B., and Braun, A.C. (1958). *Nature* 181, 1205.
Pringle, R.B., and Scheffer, R.P. (1966). *Phytopathology* 56, 1149.
Pringle, R.B., and Scheffer, R.P. (1967). *Phytopathology* 57, 1169.
Quinby, J.R., and Karper, R.E. (1949). *Agronomy J.* 41, 118.
Reiss, J. (1977). *Z. Pflanzenphysiol.* 82, 446.
Rudolph, K. (1976). *In* "Physiological Plant Pathology" (R. Heitefuss and P.H. Williams, eds.), p. 270. Springer-Verlag, Berlin and New York.
Samaddar, K.R., and Scheffer, R.P. (1968). *Plant Physiol.* 43, 21.
Sassa, T., Hayakari, S., Ikeda, M., and Miura, Y. (1971). *Agric. Biol. Chem.* 35, 2130.
Sawamura, K. (1966). *Bull. Hort. Res. Sta. (Morioka), Japan, Series C*, no. 4, p. 43-59.
Schadler, D.L., and Bateman, D.F. (1975). *Phytopathology* 65, 912.
Scheffer, R.P. (1976). *In* "Physiological Plant Pathology" (R. Heitefuss and P.H. Williams, eds.), p. 247. Springer-Verlag, Berlin and New York.
Scheffer, R.P., and Briggs, S.P. (1981). *In* "Toxins in Plant Disease" (R.D. Durbin, ed.), p. 1. Academic Press, New York.
Scheffer, R.P., and Pringle, R.B. (1961). *Nature* 191, 912.
Scheffer, R.P., and Ullstrup, A.J. (1965). *Phytopathology* 55, 1037.
Scheffer, R.P., and Walker, J.C. (1953). *Phytopathology* 43, 116.

Scheffer, R.P., Nelson, R.R., and Ullstrup, A.J. (1967). *Phytopathlogy* 57, 1288.

Schipper, A.L. Jr. (1978). *Phytopathology* 68, 866.

Scott, K.J. (1976). *In* "Physiological Plant Pathology" (R. Heitefuss and P.H. Williams, eds.), p. 719. Springer-Verlag, Berlin and New York.

Sequeira, L. (1981). *In* "Plant Disease Control: Resistance and Susceptibility" (R.C. Staples and G. Toenniessen, eds.), p. 285. J. Wiley & Sons, New York.

Shain, L., and Franich, R.A. (1981). *Physiol. Plant Path.* 19, 49.

Sinden, S.L., and Durbin, R.D. (1968). *Nature* 219, 379.

Sinden, S.L., DeVay, J.E., and Backman, P.A. (1971). *Physiol. Plant Path.* 1, 199.

Smalley, E.B., and Strong, F.M. (1974). *In* "Mycotoxins" (I.F.H. Purchase, ed.), p. 199. Elsevier Publ. Co., Amsterdam.

Smidt, M., and Kosuge, T. (1978). *Physiol. Plant Path.* 13, 203.

Staskawicz, B.J., and Panopoulos, N.J. (1979). *Phytopathology* 69, 663.

Staskawicz, B.J., Panopoulos, N.J., and Hoogenraad, N.J. (1980). *J. Bact.* 142, 720.

Steele, J.A., Uchytil, T.F., Durbin, R.D., Bhatnagar, P., and Rich, D.H. (1976). *Proc. Nat. Acad. Sci. (U.S.)* 73, 2245.

Steiner, G.W., and Byther, R.S. (1971). *Phytopathology* 61, 691.

Steiner, G.W., and Strobel, G.A. (1971). *J. Biol. Chem.* 246, 4350.

Stermer, B.A., Hart, J.H., and Scheffer, R.P. (1981). *Phytopathology* 71, 906.

Stewart, W.W. (1971). *Nature* 229, 174.

Stoessl, A. (1981). *In* "Toxins in Plant Disease" (R.D. Durbin, ed.), p. 109. Academic Press, New York.

Stout, R.G., and Cleland, R.E. (1980). *Plant Physiol.* 66, 353.

Strobel, G.A. (1973). *J. Biol. Chem.* 248, 1321.

Suzuki, Y., Knoche, H.W., and Daly, J.M. (1982). *Biorganic Chemistry* 11, 300.

Takai, S. (1980). *Can. J. Botany* 58, 658.

Tanaka, S. (1933). *Mem. College Agric. Kyoto Imp. Univ.* 28, 1-31.

Tegtmeier, K.J., Daly, J.M., and Yoder, O.C. (1982). *Phytopathology* 72, 1492.

Templeton, G.E. (1972). *In* "Microbial Toxins", Vol. 8 (S. Kadis, A. Ciegler, and S.J. Ajl, eds.), p. 169. Academic Press, New York.

Turner, J.G. (1981). *Physiol. Plant Path.* 19, 57.
Turner, J.G., and Debbage, J.M. (1982). *Physiol. Plant Path.* 20, 223.
Turner, N.C. (1973). *Amer. J. Botany* 60, 717.
Uchytil, T.F., and Durbin, R.D. (1980). *Experientia* 36, 301.
Ueno, T., Nakashima, T., Hayashi, Y., and Fukami, H. (1975). *Agr. Biol. Chem.* 39, 1115.
Ullstrup, A.J. (1972). *Ann. Rev. Phytopath.* 10, 37.
Van Alfen, N.K., and Allard-Turner, V. (1979). *Plant Physiol.* 63, 1072.
Van Alfen, N.K., and McMillan, B.D. (1982). *Phytopathology* 72, 132.
Waitz, L., and Schwartz, W. (1956). *Phytopath. Z.* 26, 297.
Walton, J.D., Earle, E.D., Yoder, O.C., and Spanswick, R.M. (1979). *Plant Physiol.* 63, 806.
Walton, J.D., Earle, E.D., and Gibson, B.W. (1982). *Biochem. Biophys. Res. Comm.* 107, 785.
Wheeler, H., and Black, H.S. (1963). *Amer. J. Botany* 50, 686.
Whiteside, J.O. (1976). *Plant Dis. Rep.* 60, 326.
Wolpert, T.J., and Dunkle, L.D. (1980). *Phytopathology* 70, 872.
Woltz, S.S. (1978). *Ann. Rev. Phytopath.* 16, 403.
Woolley, D.W., Schaffner, G., and Braun, A.C. (1952). *J. Biol. Chem.* 198, 807.
Wright, D.E. (1968). *Ann. Rev. Microbiol.* 22, 269.
Yoder, O.C. (1973). *Phytopathology* 63, 1361.
Yoder, O.C. (1976). *In* "Biochemistry and Cytology of Plant-Parasite Interactions" (K. Tomiyama et al., eds.), p. 16. Elsevier Publ. Co., Amsterdam.
Yoder, O.C. (1980). *Ann. Rev. Phytopath.* 18, 103.
Yoder, O.C., and Gracen, V.E. (1975). *Phytopathology* 65, 273.
Yoder, O.C., and Scheffer, R.P. (1969). *Phytopathology* 59, 1954.
Yoder, O.C., and Scheffer, R.P. (1973). *Phytopathology* 52, 513.

2

Structural Aspects of Toxins

V. MACKO

I. Introduction

The purpose of this review is to summarize some recent studies on structural aspects of toxins rather than to provide a comprehensive review of the whole field. In particular, new accounts of toxins that have been implicated in plant disease will be discussed.

During the past 3 years, there have been a number of reviews written which deal with structural aspects of phytotoxins. Comprehensive and critical treatment of toxin chemistry including structure and biogenetic relations, chemical synthesis, and structure-activity relationships has been published in a monograph edited by Durbin (1981). General reviews on toxins, dealing in part with structural aspects, are those by Yoder (1980), Strobel (1982), and Durbin (1983). In addition, there are review articles that focus on toxins of *Alternaria* (Nishimura and Kohmoto, 1983; Ueno *et al.*, 1982, 1983), *Helminthosporia* (Daly, 1982), host-selective toxins (Daly and Knoche, 1982; Daly, 1983), and on bacterial toxins (Durbin, 1982).

The emphasis here will be on host-selective or host-specific toxins (Scheffer, 1976; Wheeler and Luke, 1963) (HST) because a "quantum leap" in our knowledge of the chemistry of this elusive group of toxins has occurred recently. As if in response to Mitchell's plea (Mitchell, 1981) that "proposed structures should be confirmed or rejected, not simply ignored", there was an upsurge of activity, and new state-of-the-art purification and

TOXINS AND PLANT PATHOGENESIS
ISBN 0 12 200780 8

analytical techniques have been applied to some old and newer problems. As a result, the structures of several host-selective toxins have been elucidated in the last two years. An additional reason for this emphasis is the importance of HST's in disease. HST's are known to be disease determinants, molecular agents of specificity, inhibitors of gene specific plant metabolites, and as such they provide selective tools for tissue culture studies. Their use greatly simplifies the biochemical studies undertaken to develop an understanding of the molecular basis of the plant disease process. A prerequisite to such studies is a complete knowledge of the structure of each toxin.

Past studies of precise toxin mechanisms of action often represented false starts. For almost three decades, many of these studies of HST have been conducted of necessity with impure preparations of unknown structure. Thus investigations into the mode of action of these toxins have been severely hampered, and definitive interpretation of existing data has been exceedingly difficult. It has been virtually impossible to compare results from different laboratories, since they were obtained using poorly defined toxin preparations. There is now a reason for optimism. The structures of six host-selective toxins are reasonably well understood, and elucidation of the structure of several other toxins of this elusive category is near completion, thus opening the door to mode of action studies. An overview of the current state of structural studies is presented in Table 1, and a detailed discussion of HST's follows later. Table 2 summarizes the nomenclature of those *Helminthosporium* species now reclassified as *Bipolaris* (Alcorn, 1983). The *Alternaria* pathogens listed in Table 1 are now considered to be pathotypes of the species *Alternaria alternata* (Fries) Keissler (cf. Nishimura and Kohmoto, 1983).

Besides host-selective toxins, many other toxins produced by plant pathogens have been described, but not all of these substances are important in disease development. Criteria for the pathological significance of toxins have been evaluated by Yoder (1980). Several well characterized bacterial toxins have an important role in disease, but they are not host-selective.

A wide range of non host-selective toxins of fungal origin have also been described, and a great number of detailed structural characterizations performed (Stoessl, 1981), but most reports lack information on toxin

TABLE 1. Host-Selective Toxins Produced by Plant Pathogenic Fungi.

Fungus (toxin)	Chemical nature of the toxin(s)	Reference[a]
Helminthosporium sacchari (HS-toxin)	sesquiterpene glycosides	Macko et al. (1981a, b; 1982a; 1983a)
H. maydis race T (HMT- or T-toxin)	linear polyketols	Kono and Daly (1979) Kono *et al.* (1980a, 1980b)
H. carbonum race 1 (HC-toxin)	cyclic tetrapeptide	Liesh *et al.* (1982); Walton *et al.* (1982); Gross *et al.* (1982); Pope *et al.* (1983a)
H. victoriae (HV-toxin)	unknown	
Phyllosticta maydis (PM-toxin)	similar, but not identical to T-toxin; structure determination in progress	Danko *et al.* (1983)
Periconia circinata (PC-toxin)	unknown; structure determination in progress	Wolpert and Dunkle (1980)
Alternaria mali (AM-toxin)	cyclic tetrapeptides	Okuno *et al.* (1974a, b; 1975); Ueno *et al.* (1975a,b,c)
A. kikuchiana (AK-toxin)	esters of epoxydecatrienoic acid	Nakashima *et al.* (1982)
A. alternata f.sp. *lycopersici* (AAl-toxin)	dimethylheptadecapentol esters of propanetricarboxylic acid	Bottini and Gilchrist (1981); Bottini *et al.* (1981)

TABLE 1. (continued)

Fungus (toxin)	Chemical nature of the toxin(s)	Reference[a]
A. citri-rough lemon isolate (ACR-toxin)	hydrocarbon with carbonyl and hydroxyl groups (M.W. 292); structure determination in progress	Gardner *et al.* (1982)
A. citri-tangerine isolate (ACT-toxin)	unknown	Kohmoto *et al.* (1979)
A. fragariae (AF-toxin)	unknown	

[a]References are those which report structure determination of the toxins. References to early work on purification of toxins of unknown chemical nature can be found in Kono *et al.* (1981), Daly and Knoche (1982) and Nishimura and Kohmoto (1983).

involvement in disease. Such toxins will be discussed only briefly in this account.

II. Methodology

Recently, a clear trend in the approaches to structural characterization of toxins, especially host selective toxins, has emerged. Methods used in structural studies of these compounds include the bioassay, extraction and separation techniques, analytical characterization, and eventually organic synthesis. The rapid development of these methods, with consequent high levels of specialization in each area, makes it almost impossible for a single team to tackle the whole complex procedure. The successful solution of such problems often necessitates a multidisciplinary approach. Examples of such multiauthor teams can be found in this chapter in the accounts on HC-toxin, T-toxin, HS-toxin and others.

TABLE 2. Synonyms for *Helminthosporium* - *Bipolaris* species producing host-selective toxins.[a]

Anamorph (conidial state)	Teleomorph (ascus state)
Helminthosporium sacchari (Van Breda de Haan) Butler *Bipolaris sacchari* (Butl.) Shoem. *Drechslera sacchari* (Butler) Subram. and Jain	Unknown
Helminthosporium maydis Nisik. *Bipolaris maydis* (Nisikado) Shoem. *Drechslera maydis* (Nisik.) Subram. and Jain	*Cochliobolus heterostrophus* Drechsler
Helminthosporium carbonum Ullstrup *Bipolaris zeicola* (Stout) Shoem. *Helminthosporium zeicola* Stout *Drechslera zeicola* (Stout) Subram. and Jain	*Cochliobolus carbonum* Nelson
Helminthosporium victoriae Meehan and Murphy *Bipolaris victoriae* (Meehan and Murphy) Shoem. *Drechslera victoriae* (Meehan and Murphy) Subram. and Jain	*Cochliobolus victoriae* Nelson

[a]Most recent revision is that by Alcorn (1983).

A. *Biological Assay*

A critical and complete account of requirements for an acceptable bioassay in the isolation and subsequent structural elucidation of toxins was recently provided by Yoder (1981). However, the search for non-toxic compounds that are closely related to the toxin may require an indirect assay. In one case, isolation of the HS-toxins has facilitated the development of a rational approach to the isolation of their lower homologs. These compounds are not only non-toxic to sugarcane tissue, but they can act as protectants. Pre-treatment of susceptible sugarcane tissue with solutions of the isolated lower homologs of HS-toxins prevents the toxic action of HS-toxin (Livingston and Scheffer, 1981a, Duvick *et al.*, 1983). Do other pathogens produce similar protectant compounds? It should be possible to devise a bioassay to screen for such protectant compounds in fractionated extracts of culture filtrates of fungi for which the toxins have been well characterized.

B. *Extraction and Purification Techniques*

Conventional solvent extraction is not efficient or is sometimes unsuitable for isolation of some of the more polar toxins, and is now being replaced to some extent by solid phase extraction techniques. Ambersorb XE-348 was used for HS-toxins (Macko *et al.*, 1981b), XAD-2 for AK-toxins (Nakashima *et al.*, 1982) and activated carbon for T-toxin (Kono and Daly, 1979).

C. *Chromatographic Techniques*

Isolation of a single chemical species is essential for degradation studies by chemical methods and makes analysis by spectroscopic means more reliable and straightforward. Consequently, various types of chromatography are necessary to achieve this high degree of purity. The purity of a sample defined as a single spot (peak) using several chromatography systems may be insufficient, and NMR spectra sometimes reveal the presence of more than one chemical species.

There is, however, a great variety of chromatographic supports available, and most of the separation problems can now be solved by high performance liquid chromatography (HPLC). A wide range of HPLC packings are available, representing all the modes of separation that are known for column chromatography. Reverse phase HPLC, with octadecylsilane (ODS or C_{18}) as the most popular column packing material, has proved to be the most adaptable and reliable method of separation of scores of biologically active compounds, including toxins. It served, for example, as a final purification step for HC-toxin (Walton *et al.*, 1982; Ciuffetti *et al.*, 1983; Pope *et al.*, 1983a), and the separation of the three isomeric HS-toxins was also accomplished by reverse phase HPLC (Macko *et al.*, 1981b).

A less well known technique that often gives different and unusual selectivities, is *counter current chromatography* (CCC), a method that totally eliminates the use of solid supports. The name was derived from two classical partition methods, *counter current* distribution (CCD) and liquid *chromatography* (LC). CCC has inherited advantages from both parent methods. It yields highly efficient separations comparable to LC, while retaining a good sample recovery rate, high purity of fractions and reproducibility, all of which are CCD advantages (Ito, 1981). Basic principles of CCC and its various forms such as droplet-, locular-, and coil-planet- centrifuge-CCC were reviewed by Ito (1981).

Droplet CCC was effectively used in the purification of HS- and AK-toxins (Macko *et al.*, 1981b; Nakashima *et al.*, 1982). Excellent and relatively fast separation of the homologous glycosides from *H. sacchari* can be obtained with Ito's new multi-layer coil CCC apparatus (Kratky, Macko *et al.*, unpublished). This recent CCC technique utilizes a "column" 160 meters long, 1.6 mm I.D., without a solid support. The combination of HPLC and CCC as complementary methods provides separation power that was never available in the past.

D. *Analytical Techniques*

The most significant recent analytical innovation has occurred in mass spectrometry. In the past, mass spectrometry of many biologically interesting compounds, such as toxins, was problematic. These compounds are often polar, multifunctional and nonvolatile, and thus unsuitable for electron impact or conventional chemical ionization mass spectrometry. New techniques such as fast atom bombardment mass spectrometry (FAB-MS) and field desorption (FD-MS) usually give a molecular or pseudomolecular ion so that the molecular weight and in most cases empirical formula can be clearly established.

E. *Literature Search*

An important contribution to structural studies of toxins can be made by a literature search. For example, Liesch *et al.* (1982) used a computer-based natural products structure recognition system. This allowed recognition of a net structural similarity of HC-toxin with the previously identified chlamydocin and Cyl-2 and led to the suggestion that the unusual amino acid, 2-amino-8-oxo-9,10-epoxydecanoic acid (AOE), is an integral part of the HC-toxin. The computer system used and developed by Liesch and Albers-Schonberg or that offered by Chemical Abstracts (American Chemical Society) should serve not only in aiding structural elucidation, but should also help in finding connections between unrelated fields. The similarity of chlamydocin, a potential anticancer agent, and the HC-toxin was uncovered in this manner (Liesch *et al.*, 1982).

III. Host-selective Toxins of Fungal Origin

A. HS-Toxin

1. Structure of the Three Isomeric HS-Toxins. Reports on structures of the host-selective toxins from the fungus *Helminthosporium sacchari*, the causal organism of eyespot disease of sugarcane, have presented conflicting results which are summarized in Table 3. The structure proposed by Steiner and Strobel (1971) was shown to be incorrect (Beier, 1980; Kono *et al.*, 1981a). The main problem with other tentative structural proposals for the toxin as a diglycoside (Beier, 1980), a pentaglycoside (Livingston and Scheffer, 1981b) or as a tetraglycoside (Beier *et al.*, 1982) was that pure chemical species were not available and apparently a mixture of the isomers was analyzed. Only after pure toxins were isolated was the structure of the three isomers established. This involved a collaborative effort of Prof. Arigoni and his associates in Switzerland and workers at Boyce Thompson Institute at Cornell University in the U.S.A. (Macko *et al.*, 1981a, b, 1982a, 1983a) and these results are summarized below.

Several modes of mass spectrometry were used to analyze the HS-toxins. These included fast atom bombardment, field desorption, chemical ionization/desorption mass spectrometry with ammonia as reagent gas, negative chemical ionization with Freon as reagent gas, and high and low resolution electron impact mass spectrometry. Molecular ions for the three toxins were identical, suggesting a molecular weight of 884 for each isomer. Fragment ions corresponded to a sequential loss of four monosaccharide units from the molecular ion. Fragments 200 and 218 were attributed to the sequential elimination of water from an aglycone of molecular weight 236. Analysis of hydrolysis products of the toxin by gas and thin layer chromatography indicated the

TABLE 3. Proposed Structures of Host-Selective Toxins from *Helminthosporium sacchari*

Structure	Mol. weight	Reference
2-hydroxycyclopropyl-α-D-galactopyranoside (helminthosporoside)	236	Steiner and Strobel (1971)
Two galactose units linked to $C_{15}H_{22}O_2$	558	Beier (1980)
Five galactose units linked to $C_{15}H_{22}$ moiety	1028	Livingston and Scheffer (1981b)
Three isomeric glycosidic components, containing four galactose units linked to a sesquiterpene aglycone $C_{15}H_{24}O_2$	884	Macko, Arigoni and Acklin (1981a); Macko, Goodfriend, Wachs, Renwick, Acklin and Arigoni (1981b)
Two 5-O-β-galacto-furanosyl-β-galacto-furanose units attached to $C_{15}H_{22}$ moiety	884	Beier, Mundy and Strobel (1982)
Complete structure for the three isomers as in Fig. 1.	884	Macko, Acklin and Arigoni (1982a); Macko, Acklin, Hildenbrand, Weibel and Arigoni (1983a)

presence of only one monosaccharide which was identical to galactose.

From these data it was concluded that the three isomeric toxins have the composition $C_{39}H_{64}O_{22}$ corresponding to a four-fold galactosidation of a sesquiterpene aglycone, $C_{15}H_{24}O_2$. Extended ^{1}H and ^{13}C NMR investigations confirmed this conclusion and revealed the detailed structure of the three isomers, now referred to as HS-toxins A, B and C (Fig. 1).

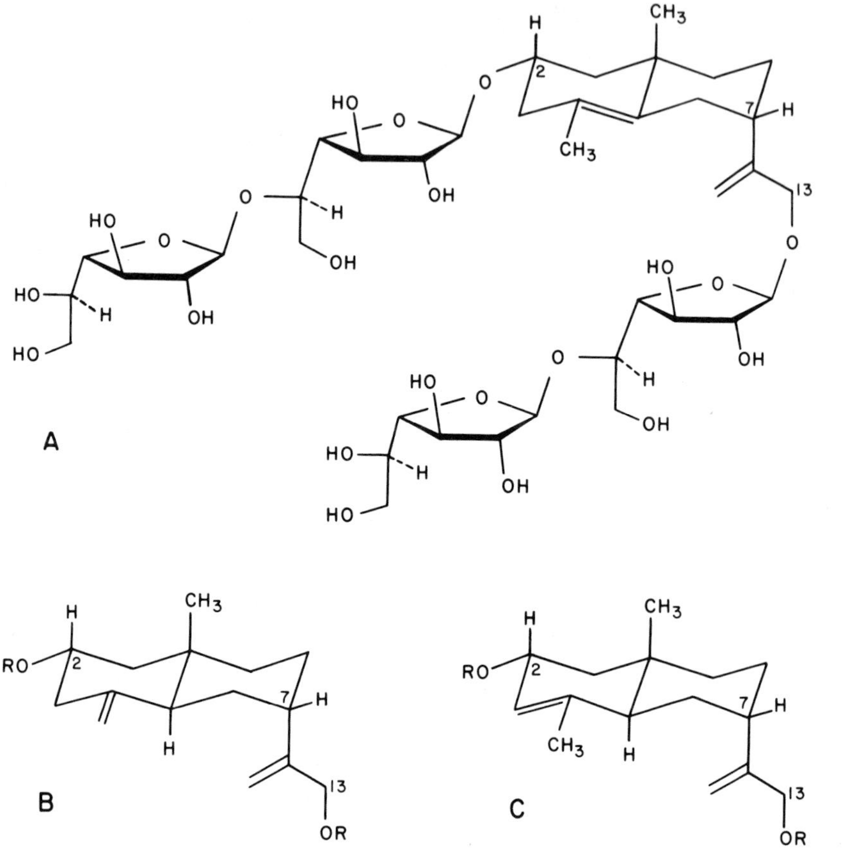

A

B

C

R=5-O-(β-galactofuranosyl)-β-galactofuranosyl-

FIGURE 1. Structures of the three isomeric toxins from *Helminthosporium sacchari;* A, B and C denote the three types of sesquiterpene aglycones (Macko *et al.*, 1982a, 1983a).

^{13}C NMR spectra in ^{1}H noise decoupled and the ^{1}H off-resonance modes were consistent with the presence of 39 carbon atoms. The ^{13}C-resonances for galactose residues showed that the four units must be arranged in two groups with the aglycone in the middle. This proposal was supported by the pattern of resonances in the anomeric region of ^{1}H spectra. Additional evidence for this arrangement was obtained from comparison with the ^{13}C and ^{1}H NMR spectra of a synthetic benzyl-5-O-(β-galactofuranosyl)-β-galactofuranoside. Independent and conclusive evidence for the central positioning was obtained eventually through analysis of a set of co-occurring lower homologs described in the next section (III.A.2).

The ^{13}C-NMR spectra of the aglycone component of each of the three toxins revealed the presence of four sp^2- and two oxygen carrying sp^3-carbon atoms. Since according to the IR-spectrum the toxins did not contain carbonyl groups, it followed that all three aglycones must possess a doubly unsaturated bicarbocyclic framework. Comparison of ^{1}H- and ^{13}C-NMR data of the three toxins clearly showed that the three isomers differed only in the position of one double bond.

Structural details of the sesquiterpene moieties were deduced from 300 and 600 MHz ^{1}H NMR spectra. Relative stereochemistry was deduced partly from coupling constants and partly from chemical shift differences in the three isomers. Conclusions were backed by Nuclear Overhauser Effect enhancements. The terpenoid origin of the three aglycone components was verified in feeding experiments with 5-^{13}C mevalonate and their absolute configuration assigned by analogy to related sesquiterpenes from other *Helminthosporium* species (Dorn and Arigoni, 1974; Arigoni, 1975).

2. Lower Homologs of HS-Toxin, Alias "Toxoids" or "Noxins". During the isolation of toxins from culture filtrates of *Helminthosporium sacchari*, it became apparent that closely related and less polar but non-toxic compounds, apparently lower homologs of HS-toxins, were also present in the extracts (Livingston and Scheffer 1981a,b; Macko *et al.* 1981a, 1982b).

Lower homologs were first detected by Livingston and Scheffer (1981a) who indicated the presence of four compounds, called by them noxins, that differed from toxins and from each other by the number of galactose units. Later, three similar compounds were referred to as toxoids and this now appears to be the preferred designation (Livingston and Scheffer, 1982). More recently these authors demonstrated by acid hydrolysis that HS-toxin contains four galactose units and that the three toxoids contain three, two and one units of galactose respectively (Livingston and Scheffer, 1983).

In a joint effort, the Boyce Thompson Institute and Zürich groups succeeded in detecting and in characterizing the structures of 21 new glycosidic compounds, lower homologs of HS-toxins, from culture filtrates of *H. sacchari* (Macko *et al.*, 1982b, 1983b). On the basis of their molecular weights, obtained by fast atom bombardment mass spectrometry, they can be divided into three groups. In the first group, six compounds each had a molecular weight of 722. In the second group, nine compounds each had a molecular weight 560, and the last group consisted of an additional 6 compounds of molecular weight 398. The observed multiplicity and distribution of compounds made it plausible that they represent the outcome of a (formal) sequential random loss of galactofuranose units from the three isomeric HS-toxins A, B and C. This working hypothesis was fully verified by a detailed analysis of the NMR spectra of the 21 compounds, which allowed derivation of the structures summarized in Table 4. In this table and in subsequent discussion the HS-toxins A, B and C are designated as $A_{2,2}$, $B_{2,2}$ and $C_{2,2}$ respectively. A loss of galactofuranose units from either side of their molecules is shown as $A_{2,1}$, $A_{1,2}$, $B_{2,1}$, $B_{1,2}$, etc.

A direct correlation between the toxins and their lower homologs was achieved with the help of an enzyme partially purified from culture filtrates of *H. sacchari* and capable of hydrolyzing a synthetic substrate, benzyl-5-0-(β-galactofuranosyl)-β-galactofuranoside. Thus, treatment of HS-toxin C (alternative designation $C_{2,2}$) with this preparation resulted in the appearance of $C_{2,1}$, $C_{1,2}$, $C_{1,1}$, $C_{0,2}$, $C_{2,0}$, $C_{1,0}$ and $C_{0,1}$ in the reaction mixture (Fig. 2) (Schröter *et al.*, 1982). Similar results were obtained with β-D-galactofuranosidase isolated from *Penicilium charlesii* (Rietschel-Berst *et al.*, 1977) (this enzyme preparation was obtained from Dr. J.E. Gander).

Livingston and Scheffer (1983) working with a mixture of the three isomeric toxins and without any separation of isomers within the group of the generated lower homologs have obtained similar results, namely that the enzyme converted the toxin to the three lower homologs (toxoids) *in vitro*. Nevertheless, both sets of results are mutually supportive and provide convincing evidence for the presence of lower homologs of HS-toxins in the culture filtrates of *H. sacchari*.

Low levels of β-galactofuranosidase activity were found in cultures and higher levels occurred in mycelium of *H. sacchari* (Livingston and Scheffer, 1983). Similar enzyme activity was found in cultures of *H. victoriae*, *H. maydis*,

TABLE 4. Host Selective Toxins and Their Non-Toxic Lower Homologs Produced by *Helminthosporium sacchari*.[a]

Tetraglycosides (HS-toxins A, B and C) M.W. 884	$A_{2,2}$	$B_{2,2}$	$C_{2,2}$
Triglycosides, M.W. 722	$A_{2,1}$	$B_{2,1}$	$C_{2,1}$
	$A_{1,2}$	$B_{1,2}$	$C_{1,2}$
Diglycosides, M.W. 560	$A_{1,1}$	$B_{1,1}$	$C_{1,1}$
	$A_{2,0}$	$B_{2,0}$	$C_{2,0}$
	$A_{0,2}$	$B_{0,2}$	$C_{0,2}$
Monoglycosides, M.W. 398	$A_{1,0}$	$B_{1,0}$	$C_{1,0}$
	$A_{0,1}$	$B_{0,1}$	$C_{0,1}$

[a]A, B and C denote the structures of the aglycone moieties (cf. Fig. 1) and the appended suffixes indicate the number of β-D-galactofuranose units in the sugar extensions at C-2 and C-13, respectively. HS-toxins have been denominated accordingly to underline their kinship to the lower homologs.

H. carbonum, and *H. turcicum*, but not in *Fusarium oxysporum* and *Cladosporium cucumerinum* (Livingston and Scheffer, 1982, 1983).

These results suggest that the lower homologs found in the culture filtrates are degradation products from, rather than biosynthetic precursors of the toxins. In addition, the sensitivity of the enzyme provides a direct clue for the D- and β-configurations of the galactose units. Moreover, the generation of the isomeric compounds $C_{2,1}$ and $C_{1,2}$ verifies that their toxic precursor is correctly formulated as $C_{2,2}$ rather than $C_{3,1}$, $C_{1,3}$, $C_{0,4}$ or $C_{4,0}$. The two latter possibilities were already excluded by the

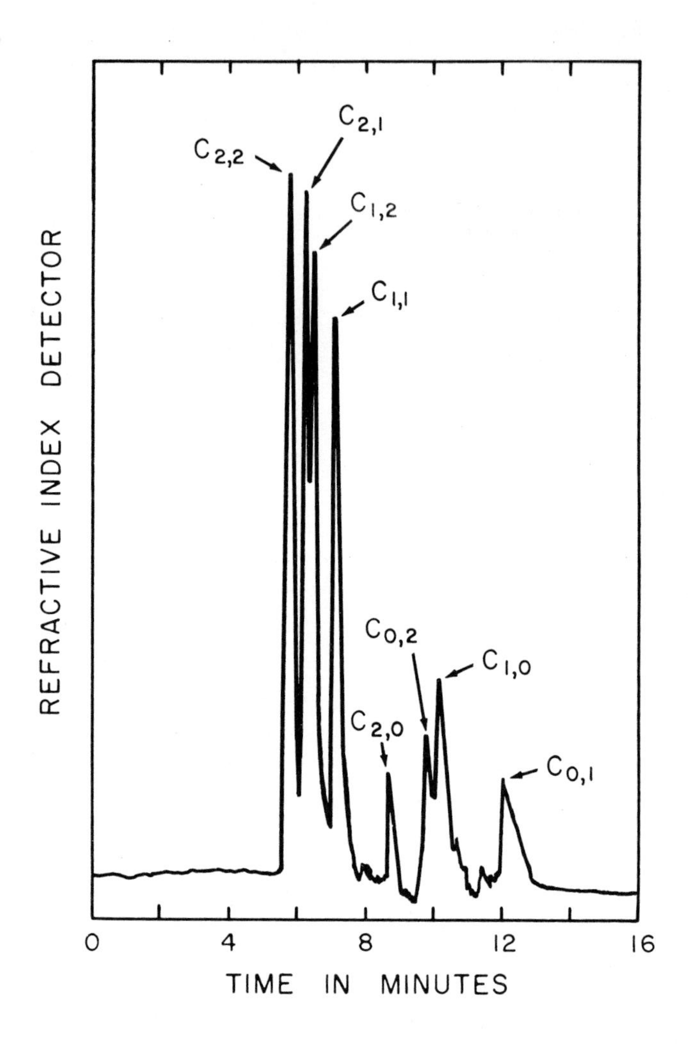

FIGURE 2. HPLC of products formed from HS-toxin C, which is designated in the chromatogram as $C_{2,2}$, after it was incubated *in vitro* with purified β-galactofuranosidase from culture filtrate of *Helminthosporium sacchari*. Column: Reverse phase C_{18}. Solvent: 17% CH_3CN in H_2O at 1 ml/min. (Schröter *et al.*, 1982).

observed ratio of 1:1 of external to internal galactose units in the toxins (Beier *et al.*, 1982; Macko *et al.*, 1983a).

3. Penta- and Hexaglycosides and the "Latent" Form of HS-Toxins. Further collaborative work of Weibel, Acklin and Arigoni in Zürich and Macko and Schröter at Boyce Thompson Institute (unpublished) showed the presence of a number of pentaglycosides (M.W. 1046) as well as three non-toxic compounds similar to HS-toxins in culture filtrates of *H. sacchari*. The molecular weight of each of the three non-toxic compounds was determined to be 1208. This indicated a HS-toxin moiety with the attachment of two additional hexose units (884 + 2x162 = 1208). Comparison of the NMR data for the three isomers with those for the three corresponding toxins indicated the additional presence of α-glucopyranosyl units linked to positions 2 of the external galactofuranose units. The structure of the "latent" form of HS-toxin C is shown in Fig. 3. In accordance with expectation, incubation of this compound with a commercial α-glucosidase yielded the known HS-toxin C; β-glucosidase had no effect. Further studies are needed to ascertain the biological significance of these congeners of HS-toxins.

4. Biological Activity of HS-Toxins and Their Congeners. Four assays are available to date to ascertain the biological activity of HS-toxins and related compounds.

a. Leaf-drop assay. When tested in a leaf-drop assay, which was introduced for HS-toxin by Steiner and Byther (1971), HS-toxins A, B and C showed similar activity, producing short water-logged runners at minimum levels of $2x10^{-11}$ moles (17.7 ng) for the susceptible clone Co-453 and $2x10^{-7}$ moles (177 μg) for the resistant CP 76-1343 clone (Macko *et al.*, 1981a,b). This assay has its advantages when testing large numbers of chromatographic fractions for activity, but it is much less suitable for quantitative work because it gives poor dosage response.

b. Electrolyte leakage assay. Assays based on toxin-induced leakage of electrolyte from tissues have been useful in studies of several host-selective toxins. Scheffer and Livingston (1980), working with HS-toxin preparations of various purity, reported a minimum concentration of 10 ng/ml that induced leakage of electrolytes from leaf disks of susceptible clone Co 453, whereas 100 μg/ml had no effect on an insensitive clone, HS2-4610.

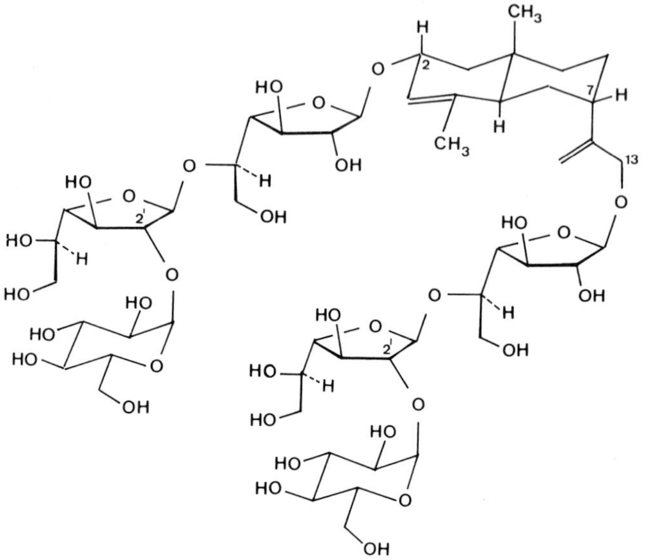

FIGURE 3. "Latent form" of HS-toxin C.

c. Inhibition of dark CO_2 fixation assay. This assay is rapid, sensitive and gives a good dosage-response curve for T-toxin (Daly and Barna, 1980; Suzuki *et al.*, 1982b). With this assay Duvick *et al.* (1983) found that HS-toxin C caused 50% inhibition at $7x10^{-7}$ M whereas toxin isomers B and A were active at somewhat lower concentrations of $3x10^{-6}$ and $4x10^{-6}$ M respectively. Preincubation (15 min) of susceptible sugarcane tissue with the lower homologs abolished the inhibitory effect of toxin. "Latent" toxin was slightly inhibitory at 10^{-4} M but the presence of α-glucosidase in the tissue that could "activate" this form of toxin was not tested.

d. Depolarization of membrane potential. Highly specialized equipment is a prerequisite for this method (Novacky and Ullrich-Eberius, 1982). Schröter *et al.* (1983) found that the energy-dependent component of membrane potential in cells of susceptible sugarcane is quickly diminished after application of HS-toxin. The lag time between the toxin application and a first sign of membrane depolarization is 4-15 min dependent on toxin concentration. The minimum toxin concentration causing membrane effect was as low as 5×10^{-8} M. Treatments with lower homologs prevented depolarization induced by HS-toxin.

B. T-Toxin

1. Structure of the T-Toxin. The phytopathogenic fungus *Helminthosporium maydis*, race T, produces a complex of chemically analogous host-selective toxins, that have a specific blighting effect on leaves of sensitive corn (Kono and Daly, 1979; Kono *et al.*, 1980 a,b). T-toxins are highly toxic (10^{-8} - 10^{-9} M) towards cells and organelles of susceptible corn possessing Texas male sterile (Tms) cytoplasm. The toxin complex was found by thin layer chromatography on silica-gel to be a mixture of eight to ten linear polyketols varying from C_{35} to C_{45} in length, with each component possessing apparently identical specific toxicity toward Tms corn. The C_{39} and C_{41} components, comprising 60-90% of the native toxin, have been characterized (Fig. 4).

The hydroxyl-bearing carbons (n) of the toxins are chiral centers, so that 2^n isomers are possible for each of these compounds. Thus there is ample scope for stereoisomerism, and some of the structural details have been proposed. Comparison of one of the degradation products of these toxins, 3-hydroxyoctanoic acid, with authentic 3(R)-hydroxyoctanoic acid suggested that the configuration at C-6 carbon is R. The hydroxyl groups at C-6 and C-8 appear to be *cis* by comparison with synthetic *cis* and *trans* dihydroxypentadecane-6-one and their phenyl boronates. Consequently the configuration on C-8 carbon should be R (Kono *et al.*, 1981b).

Evidence for the validity of the isolation procedure for T-toxin was obtained in a study of cultures of near-isogenic isolates of *H. maydis* races T and O. When segregated monogenically for race differences, the T-toxin complex was produced by race T but not by race O (Tegtmeier *et al.*, 1982). Thus in these crosses a single gene appears to control the production of the T-toxin complex, although several other loci may be involved (Tegtmeier *et al.*, 1982). However, as the authors noted, these results neither support nor eliminate the possibility that there is a block in a normal metabolic pathway of race T leading to accumulation of metabolic product (T-toxin), nor do they exclude the possibility that race O synthesizes T-toxin but has additional enzymes capable of further metabolizing it. A

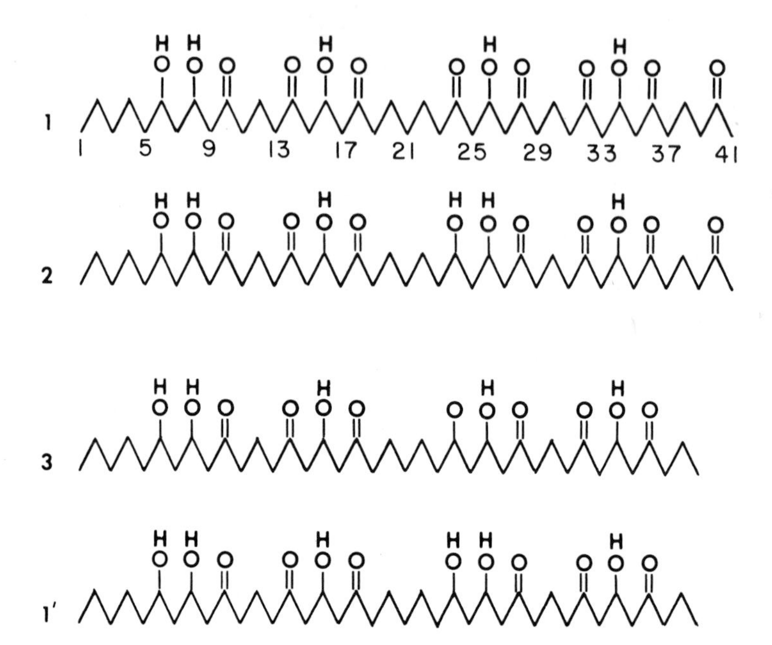

FIGURE 4. Major components of the host-selective toxin complex produced by *Helminthosporium maydis* race T. (Kono and Daly, 1979; Kono *et al.*, 1980a,b; 1981b,c).

need for knowledge of biosynthesis and metabolism of T-toxins is apparent (Tegtmeier *et al.*, 1982).

2. Synthetic Analogs of T-Toxin. A significant advance in the structural chemistry of T-toxins was made when fourteen analogs of these toxins were synthesized. Grignard additions of aldehyde intermediates to di(bromomagnesium) alkanes resulted in shorter versions (C_{15} to C_{26}) of native toxins containing the β-polyketol functions, which appear to account for the specificity and very high activity of T toxin (Suzuki *et al.* 1982a).

Inhibition of dark CO_2 fixation by susceptible corn leaves was used as a bioassay to compare the relative toxicity of synthetic analogs with that of the native toxin. As shown in Fig. 5, analogs with C_{15}, C_{25} and C_{26} chain lengths and 1,5-dioxo-3-hydroxy functions were only slightly less toxic (2-6 x 10^{-7} M) than the individual components (C_{35}-C_{45} chain lengths) of the native T toxin (3 x 10^{-8} M) (Suzuki *et al.* 1982b). Like the native toxin, analogs were

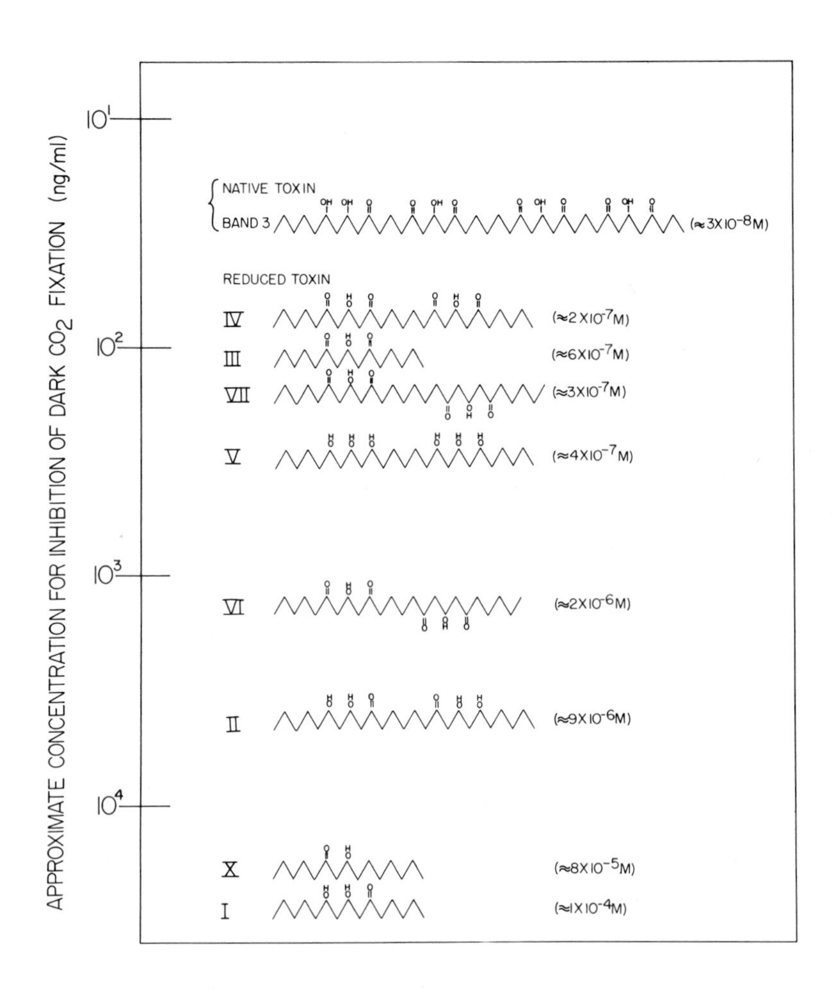

FIGURE 5. Relative effectiveness of synthetic race T-toxin analogs in inhibiting dark CO_2 fixation in leaves of susceptible Tms corn. The concentrations represent the amounts required to obtain 25% inhibition of fixation. Band 3 toxin is a purified component from the mixture of host-selective polyketols comprising native T-toxin. Reduced toxin is a series of linear polyalcohols prepared by reducing native toxin with sodium borohydride, and IV, III, VII, etc. are synthetic analogs of the native toxin complex. Reproduced with permission from Suzuki *et al.*, 1982b.

host-selective. The analogs did not inhibit dark CO_2 fixation in leaf tissue of resistant corn at concentrations 10^2 or 10^3 times greater than those effective with susceptible corn. A very close (C_{41}) analog of T-toxin, with almost the same distribution pattern of carbonyl and hydroxyl groups as T-toxin 3 (Fig. 4) was recently synthesized. Biological activity of this synthetic compound in the dark CO_2 fixation assay was indistinguishable from that of the native T-toxin complex (Suzuki *et al.*, 1983). These findings support the validity of the structure previously proposed for native T-toxin.

Some host-selectivity may occur with chemically distinct compounds (Tegtmeier *et al.*, 1982). Reduced toxin, a linear polyalcohol, possesses high activity and specificity without the carbonyl groups that are present in the native toxins and their synthetic analogs (Suzuki *et al.*, 1982b). The insecticide methomyl is structurally quite different from T-toxin, but has the same host selectivity, although at much higher concentrations (10^{-3} M), and the corn pathogen *Phyllosticta maydis* also produces a toxin of unknown composition with the same host-selectivity (Comstock *et al.*, 1973; Yoder, 1973). In this last case, however, it was recently found that the toxin complex from *Phyllosticta maydis* is structurally related, though not identical, to T-toxin (Danko *et al.*, 1983).

The number of chemical species in the T-toxin complex has been determined in acetone- or chloroform-insoluble precipitates, but activity also occurs in the soluble portion (Tegtmeier *et al.*, 1982). It is therefore more than likely that additional species of T-toxin will emerge as newer methods of HPLC are applied to extracts.

C. *HC-Toxin*

1. Structure of the HC-Toxin. Early isolation work and attempts to elucidate the structure of the toxin from a corn-attacking fungus, *Helminthosporium carbonum*, race 1, were performed by Pringle and Scheffer (1967) and Pringle (1970, 1971). They postulated a cyclic polypeptide ($C_{32}H_{50}N_6O_{11}$) containing proline, alanine and unknown amino acids. Recently, the isolation and structural elucidation of the HC-toxin was reported by several independent groups. Liesh *et al.* (1982) have demonstrated that the toxin is a cyclic terapeptide of molecular weight 436, consisting of alanine (2 residues), proline, and the unusual amino acid, 2-amino-8-oxo-9,10-epoxydecanoic acid (AOE). Three

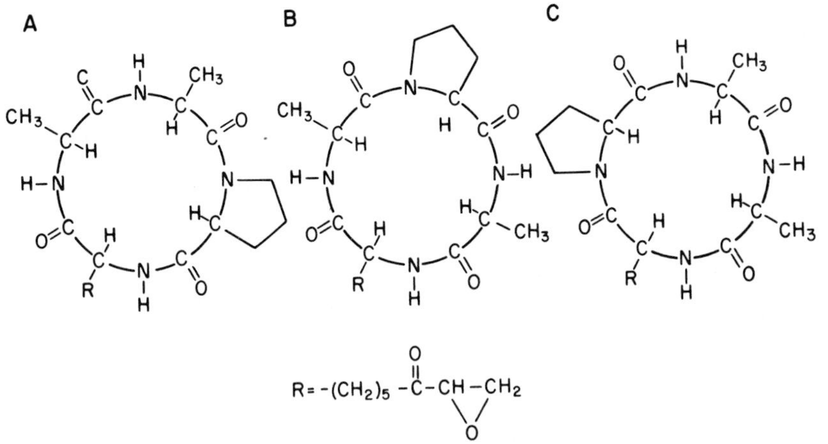

FIGURE 8. Possible amino acid sequences for the HC toxin. Leish *et al.* (1982) proposed sequence A, but there is now convincing evidence for sequence C (Walton *et al.*, 1982; Gross *et al.*, 1982; Pope *et al.*, 1983).

sequences are possible for these amino acids (Fig. 8; A, B, C). On the basis of fragmentation during EI-MS, Liesh *et al.* (1982) proposed the amino acid sequence AOE-alanine-alanine-proline (Fig. 8A). This sequence assignment is considered to be equivocal because of the possibility of rearrangements during EI-MS fragmentation (Anderegg *et al.*, 1979). Extensive studies by Walton *et al.* (1982), Gross *et al.* (1982) and Pope *et al.* (1983a) support the same amino acid composition, but produced detailed and convincing evidence for a different amino acid sequence, namely AOE-proline-alanine-alanine (Fig. 8C).

Walton *et al.* (1982) used gas chromatography-mass spectrometry as their principal method for sequencing of the HC-toxin. This included partial acid hydrolysis of the sample, methylation, trifluoroacetylation, reduction to the polyamino-alcohols and silylation of the resulting mixture of peptides, which were then identified by GC-MS (Carr *et al.* 1981). Gross *et al.* (1982) determined the sequence using fast atom bombardment mass spectrometry and mass spectrometry/mass spectrometry. A collision-induced decomposition spectrum of the protonated molecule $(M+H)^+$ m/e 437 was obtained by selecting it from one mass spectrometer and then activating by collision with helium and scanning the spectrum of products in a second, coupled, mass spectrometer. The sequence of amino acids was then deduced from the spectrum of decomposition products. Pope *et al.* (1983a) prepared a series of HC-toxin derivatives which were used to establish the identity of AOE by mass spectrometry and ^{1}H NMR. The sequence of amino acids was determined by partial acid hydrolysis and identification of the resulting dipeptides, after dansylation, by TLC. Structural assignments for the HC-toxin are quite convincing, since these groups independently reached the same conclusions using different methods. HC-toxin was thus identified as cyclo[(2-amino-9,10-epoxy-8-oxodecanoyl)-prolyl-alanyl-alanyl] (Fig. 8C).

Walton *et al.* (1982) originally proposed the carbon chiralities to be L,D,L,L, based on digestion of the amino acids with D- and L-amino acid oxidases. However, in a recent report, this proposal for the second alanine having L-configuration was retracted (Walton and Earle, 1983). Baltzer and Johnson (1983) used amino acid oxidase digestion to show that HC-toxin contains D-proline, L-alanine and D-alanine; the configuration of AOE was not determined. Kawai *et al.* (1983b) established by detailed NMR studies and amino acid oxidase experiments that the second alanine has the D- rather than the L-configuration; thus the HC-toxin is cyclo(L-AOE-D-Pro-L-Ala-D-Ala). The chirality of the epoxide group has not been assigned. HC-toxin in chloroform adopts the bis-γ-turn conformation previously found for cytostatic tetrapeptide chlamydocin (Kawai *et al.*, 1983b; Rich *et al.*, 1983). Similar results were obtained by Pope *et al.* (1983b) and Mascagni *et al.* (1983). Since the structures of HC-toxin and chlamydocin are remarkably similar, and extensive information on the chemistry of chlamydocin and its derivatives is available (Pastuszak *et al.*, 1982; Kawai *et al.*, 1983a), synthesis of HC-toxin can be expected soon.

Purified HC-toxin preparation was reported to cause 50% inhibition of susceptible root growth at 0.2 μg/ml (Ciuffetti *et al.*, 1983), 0.5 μg/ml (Walton *et al.*, 1982), and 1 μg/ml (Liesh *et al.*, 1982), in a standard bioassay. Ciuffetti *et al.* (1983) found that the range of active concentrations is apparently due to the presence of some inactivated HC-toxin in the preparations of Walton *et al.* (1982) and Liesh *et al.* (1982).

2. *Inactivation of HC-Toxin.* The amino acid, 2-amino-8-oxo-9,10-epoxydecanoic acid (AOE) has been found in three natural products; the toxins Cyl-2 (Hirota *et al.*, 1973), HC-toxin and chlamydocin (Closse and Huguenin, 1974), which is a cytostatic tetrapeptide. A similar compound,

9,10-epoxydecatrienoic acid, was recently found as a component part of the AK-toxin (see below). There is now considerable experimental evidence available that AEO is essential for the biological activity of HC-toxin and chlamydocin. Ciuffetti *et al.* (1983), in a careful study, found that storage of HC-toxin (Fig. 9A) in dilute aqueous solutions of trifluoroacetic acid resulted in hydrolysis of the oxirane (epoxy) ring of the AOE moiety of HC-toxin. The conversion product (Fig. 9B) was isolated, and complete evidence was provided for its structure. The conversion product was inactive at concentrations 100 times greater than native toxin and showed no competitive effect with it. Apparently the hydrolysis of the epoxy function destroys the interaction between the molecule and a hypothetical toxin receptor site(s).

Walton and Earle (1983), in a similar study, subjected HC-toxin to mild acid conditions (HCl and methanol) which also resulted in loss of biological activity. Mass spectrometry of the acid-degraded toxin demonstrated the presence of several components which indicated that the terminal epoxide moiety was altered (Fig. 9C–G), resulting in the toxin inactivation.

3. Similarities Between Inactivation of HC-Toxin and Chlamydocin. Chlamydocin, a related cyclic tetrapeptide from *Diheterospora chlamydospora* identified as cyclo (L-AOE-α-aminoisobutyril-L-phenylalanyl-D-propyl) (Fig. 10A) has extremely high cytostatic activity against mouse mastocytoma cells, at 0.5 μg/ml. The activity of chlamydocin *in vivo* has been limited by its rapid inactivation in whole serum by a process which appears to be enzymatically catalyzed. The mechanism by which chlamydocin is inactivated is not known, but one probable site of inactivation would be the epoxy carbonyl group found in the amino acid AOE. Chemical alteration of the epoxide or the carbonyl moiety (Fig. 10B–E) reduces the biological activity of the molecule by a factor of 1000, suggesting an important role, possibly as an alkylating agent, for the epoxy carbonyl functional group (Closse and Huguenin, 1974; Stähelin and Trippmacher, 1974).

The structural similarities between chlamydocin and HC-toxin and their dependence on the epoxide for activity suggest that they may have a similar mode of action.

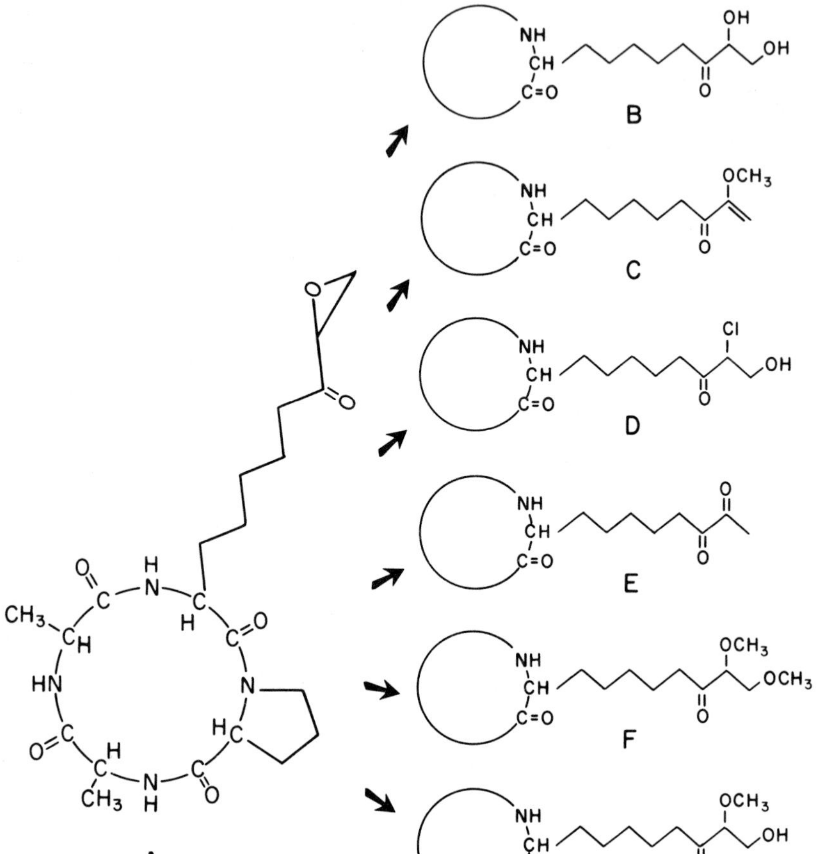

FIGURE 9. HC-toxin (A) and its inactive derivatives (B–G). Derivative B was isolated as a conversion product after storage of HC-toxin in aqueous solutions and acidic conditions (Ciuffetti *et al.*, 1983). C–G shows probable composition of compounds resulting from treatment of HC-toxin with HCl in methanol. These structures were postulated from FAB-MS of their unresolved mixture (Walton and Earle, 1983).

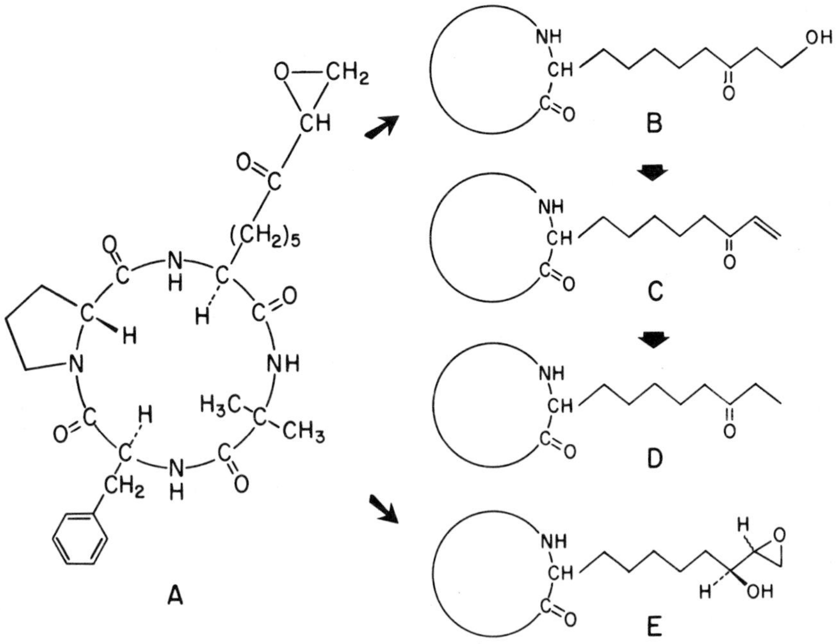

FIGURE 10. Chlamydocin (A) and its inactive derivatives (B–E). Derivatives B–D were prepared by chemical modification of A. Compound E, dihydrochlamydocin, was isolated along with A from culture filtrates of *Diheterospora chlamydospora* (Closse and Huguenin, 1974; Stähelin and Trippmacher, 1974).

D. AK-Toxin

Alternaria kikuchiana, the causal fungus of black spot disease of Japanese pear produces a host-selective toxin. A droplet of culture filtrate of the fungus produces a necrotic spot on susceptible but not on resistant cultivars. Recently, in a careful study, the structures of two host-selective toxic metabolites, AK-toxin I and II were determined. The compounds were purified from culture filtrate using repeated purification on Amberlite XAD-2, silica gel, droplet counter current chromatography, Sephadex LH-20 and recrystalization from methanol. The high resolution mass spectrum suggested the molecular formula $C_{23}H_{27}NO_6$. Three carbonyl groups in the IR spectrum were interpreted to be in an ester, an amide and a conjugated acid. The N-acetyl-β-methyl-phenylalanyl moiety in AK-toxin I was deduced from ^{1}H NMR and mass spectra. Another structural moiety of AK-toxin I was found to be 9,10-epoxy-8-hydroxy-9-methyl-2E,4Z,6E-decatrienoic acid by derivatizing and subsequent analysis of the products as an ester and as a hexahydroderivative. In addition, all of the six olefinic protons and a methine proton of AK-toxin I were reasonably assigned. The geometrical arrangements of the conjugated olefinic protons were deduced from their coupling constants and the enhancement of proton signals in NOE experiments.

The structure of AK-toxin II was derived from a similar but separate set of measurements. On the basis of these experimental results, the structures of AK-toxin I and II were determined to be 8-(α-acetylamino-β-methyl-β-phenyl-propionyloxy)-9,10-epoxy-9-methyl-2E,4Z,6E-decatrienoic acid and its β-demethyl derivative respectively (Fig. 11) (Nakashima *et al.*, 1982).

The toxins induced necrosis on leaves of a susceptible pear cultivar, Nijisseiki at 65 ng/ml, whereas there was no

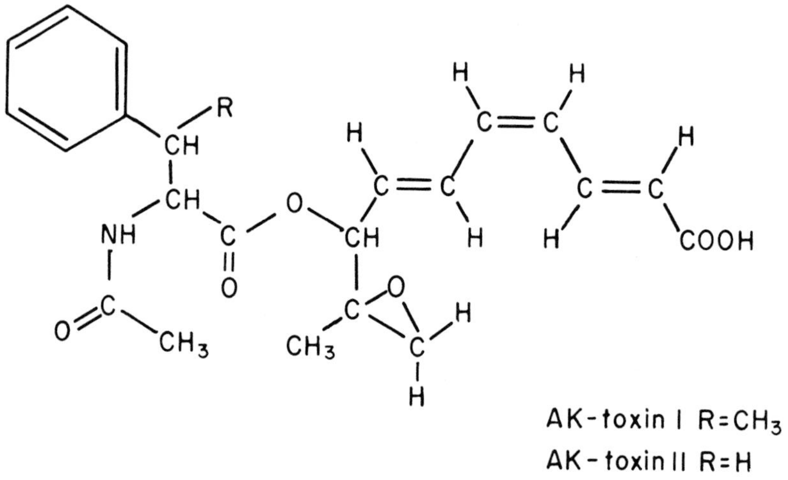

AK-toxin I R=CH_3

AK-toxin II R=H

FIGURE 11. Structure of AK-toxins (Nakashima *et al.*, 1982).

response with a resistant cultivar, Chojurs, even at a concentration of 6.5 μg/ml.

An important similarity exists between the structures of 9,10-epoxy-8-hydroxy-9-methyl-2E,4Z,6E-decatrienoic acid of the AK-toxin and AOE of the HC-toxin, Cyl-2 and chlamydocin. This raises the question whether the mode of action in these four systems is similar and whether the compounds could be cross-reactive.

E. AAl Toxin

The fungal pathogen *Alternaria alternata* f. sp. *lycopersici* causes a stem canker disease of tomato (Gilchrist and Grogan, 1976). Two toxic fractions that reproduce disease symptoms in susceptible plants at concentrations of less than 10 ng/ml have been isolated from culture filtrates of the fungus. One of these fractions (TA) was reported to consist of two esters (at C-13 and C-14) of propane-1,2,3-tricarboxylic acid and 1-amino-11,15-dimethylheptadeca-2,4,5,13,14-pentol. The second host-selectivity fraction (TB) consists of two components with the same carbon skeleton as TA, but they lack the C-5 hydroxyl and differ in stereochemistry at one or more of the chiral centers from C-11 to C-15 (Fig. 12) (Bottini *et al.*,

$$CH_3\text{-}CH_2\text{-}CH(CH_3)\text{-}CH(X)\text{-}CH(Y)\text{-}CH_2\text{-}CH(CH_3)\text{-}CH_2\text{-}CH_2\text{-}CH_2\text{-}CH_2\text{-}CH_2\text{-}CH(R)\text{-}CH(OH)\text{-}CH_2\text{-}CH(OH)\text{-}CH_2\text{-}NH_3^+$$

AAl-toxin		R	X	Y
T_A	1	OH	OH	$^-O_2C\text{-}CH_2\text{-}CH(CO_2^-)\text{-}CH_2\text{-}CO_2$
	2	OH	$^-O_2C\text{-}CH_2\text{-}CH(CO_2^-)\text{-}CH_2\text{-}CO_2$	OH
T_B	1	H	OH	$^-O_2C\text{-}CH_2\text{-}CH(CO_2^-)\text{-}CH_2\text{-}CO_2$
	2	H	$^-O_2C\text{-}CH_2\text{-}CH(CO_2^-)\text{-}CH_2\text{-}CO_2$	OH

FIGURE 12. AAl Toxins

1981a,b). Both TA and TB exhibit necrotrophic activity at less than 10 ng/ml.

A procedure for microdetermination of T_A and T_B toxins using HPLC was recently described (Silver and Gilchrist, 1982).

F. AM-Toxin

The first structural studies of host-selective toxins are those performed on AM-toxins. *Alternaria mali* causes leaf spot (*Alternaria* blotch) disease of apple and produces necrotic spots, especially on the leaves, shoots and fruits of susceptible cultivars. These host-selective symptoms are related to the presence of a host-selective toxin, a metabolite of the causal fungus. Pioneering studies by Okuno *et al.* (1974a,b and 1975) and Ueno *et al.* (1975a,b,c) resulted in almost simultaneous elucidation of the structure of AM-toxin I, and the identification of AM-toxins II and III followed (Table 5) (Ueno *et al.* 1975a,b,c). The proposed structures were confirmed by total syntheses (Lee *et al.*, 1976; Shimohigashi *et al.*, 1977; Kanmera *et al.*, 1981).

IV. Non Specific Toxins

A. Fungal

Numerous non specific or non selective fungal toxins have been described and at least 200 of these have been identified from more than 100 taxonomically different pathogens (Rudolph, 1976; Stoessl, 1981). Broadly defined, this category includes plant growth regulators, a number of mycotoxins and other biologically active compounds producing a phytotoxic effect. Many of these metabolites are toxic in bioassays, but such activity may be quite unrelated to disease. It is reasonable to assume that a systematic screening of microbial products for phytotoxicity would probably greatly increase the number of known toxins. Conversely, it should be interesting to examine toxins for their physiological effect in animal and microbial systems. However, a word of caution is in order here. Toxins should be evaluated more carefully for their role in disease. It is very likely that many microbial products assayed at a

TABLE 5. Structures and biological activity of AM-toxin I (1), II (2), III (3) and synthesized analogs (4-8). Adapted from Ueno *et al.* (1982, 1983).

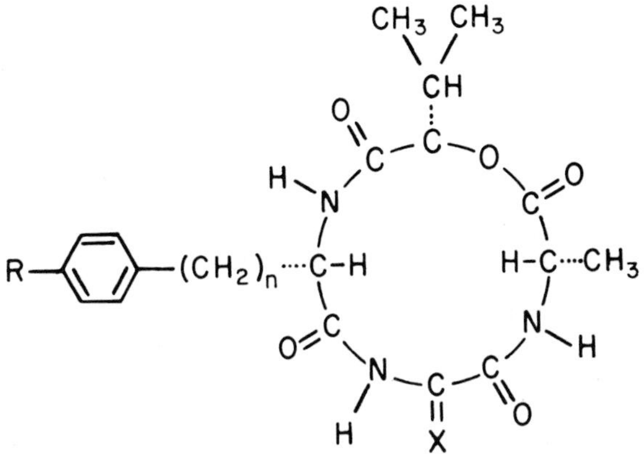

	X	R	n		Biological activity (μg/ml)
1	CH_2=	CH_3O-	3	(AM-toxin I)	0.002-0.0002
2	CH_2=	H-	3	(AM-toxin II)	0.02
3	CH_2=	HO-	3	(AM-toxin III)	0.02-0.002
4	CH_3 , H	CH_3O-	3		100
5	H, CH_3	CH_3O-	3		5-10
6	CH_2=	CH_3O-	1		20
7	CH_3 , H	CH_3O-	1		>100
8	H, CH_3	CH_3O-	1		>100

high enough concentration on a selected sensitive plant could be shown to be "phytotoxic".

Stoessl (1981) recently wrote a comprehensive review of non host-selective fungal toxins. He dealt systematically with the structure and biogenetic assignments of toxic metabolites, including compounds derived from shikimic acid, the tricarboxylic acid cycle and fatty acids. Many toxins are derived from polyketides, terpenoids and amino acid derivatives. Stoessl's review is complemented by that of Rich (1981) on chemical synthesis and of Ballio (1981) on the structure-activity relationships of toxins. These reviews cover structures of a number of classical examples of phytotoxins such as tentoxin, fusicoccin and ophiobolin, but also scores of compounds whose toxic activity is only one facet of their multiple biological activity. These latter include mycotoxins, antibiotics, growth regulators, and cytochalasins.

The large number and diversity of activities of non specific toxins can be demonstrated by the example of the trichothecenes. Since the first identification of trichothecin (Godtfredsen and Vangedal, 1964), the list of naturally occuring trichothecenes has grown to almost 60 (cf. Stoessl, 1981). These compounds have been found in *Fusarium*, *Cephalosporium*, *Myrothecium*, *Trichoderma* and *Stachybotrys* fungi (Smalley and Strong, 1974). A trichothecene triepoxide presumably originating from a soil fungus was isolated from the Brazilian shrub *Baccharis megapotamica* (Kupchan *et al.*, 1977). A chapter has even been devoted to the biosynthesis of trichothecenes (Tamm and Breitenstein, 1980). The biological activity of trichothecenes includes phytotoxic, antifungal, mycotoxicological (human and animal toxicoses), insecticidal, antiviral, antibacterial, cytostatic, cytotoxic and antileukemic effects (Smalley and Strong, 1974).

Another widely occurring group of toxins, the ophiobolins, are a class of sesterterpenes originally isolated from *Helminthosporium oryzae* (cf. Stoessl, 1981). They are also produced by several related fungal species, including *H. maydis* (Tipton *et al.*, 1977), and were recently shown to occur in *Helminthosporium sacchari* (Weibel, Acklin and Arigoni, 1982). Ophiobolin B, several closely related compounds, and other more "exotic" compounds were also isolated from culture filtrate of *H. sacchari* (Weibel *et al.*, 1982). These included (-)-4α,7β-aromadendrandiol, the enantiomer of which was previously found in extracts from soft coral (Beechan *et al.*, 1978), but was not assayed for phytotoxic activity.

These are only a few examples of the vast array of known fungal phytotoxins. Space limitations, however, preclude a systematic treatment of this broad subject of non specific fungal toxins.

B. *Bacterial*

Although species in all of the genera of phytopathogenic bacteria have been reported to produce toxins, only a few of them have been studied in detail (Durbin, 1982). In contrast to the large number of identified fungal toxins, only few bacterial toxins (phaseolotoxin, tabtoxin, rhizobitoxine, coronatine and tagetitoxin), are known in their entirety (Table 6). Tabtoxin can be hydrolyzed to tabtoxinine-β-lactam by non-specific peptidases from the plant and it is the later compound that is actually active (Uchytil and Durbin, 1980).

Since the detailed review of bacterial toxin structures by Mitchell (1981) and the more general review by Durbin (1982), only tagetitoxin has been added to the list of structurally known bacterial toxins (Mitchell and Hart, 1983). In addition, some details of the previous structural determination of phaseolotoxin have been questioned (Patil *et al.*, 1983).

1. *Tagetitoxin.* A toxin from *Pseudomonas syringae* pv. *tagetis* which causes apical chlorosis in marigolds (*Tagetes patula* L.) and zinnia has been isolated, purified (Mitchell and Durbin, 1981) and structurally characterized (Mitchell and Hart, 1983). The field desorption mass spectrum showed a molecular weight of 435; the molecular formula was deduced to be $C_{11}H_{18}NO_{13}PS$, and a cyclic hemithioketal structure (Table 6), was derived from ^{1}H- and ^{13}C-NMR spectra. Tagetitoxin produced apical chlorosis in zinnia seedlings when 10 ng was applied in 20 μl.

2. *Phaseolotoxin.* Patil *et al.* (1983) opened a discussion on the correctness of the structure proposed by Mitchell (1976) (Table 6). According to Patil *et al.* (1983), fast-atom bombardment mass spectrometry (FABMS) suggested that the molecular weight of phaseolotoxin is 531 and not 532 as required for Mitchell's structure. The molecular weight of 531 obtained by FABMS suggested to Patil *et al.* (1983) that one of the oxygen atoms in Mitchell's structure, most likely an oxygen on the phosphate group, is actually a NH. However, supporting NMR and mass spectral evidence using ^{15}N labelled toxin is not yet available.

TABLE 6. Bacterial Toxins

Organism (toxin)	Structure	Reference
Pseudomonas syringae pv. *phaseolicola* (phaseolotoxin)	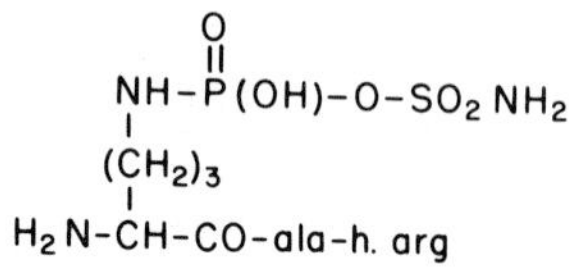 $NH-P(=O)(OH)-O-SO_2NH_2$ $(CH_2)_3$ $H_2N-CH-CO-ala-h.arg$	Mitchell (1976a,b)
[(2-serine)-phaseolotoxin]	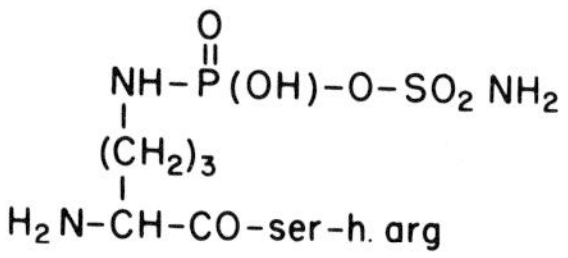 $NH-P(=O)(OH)-O-SO_2NH_2$ $(CH_2)_3$ $H_2N-CH-CO-ser-h.arg$	Mitchell and Parsons (1977)
Pseudomonas syringae pv. *tabaci* (tabtoxin)	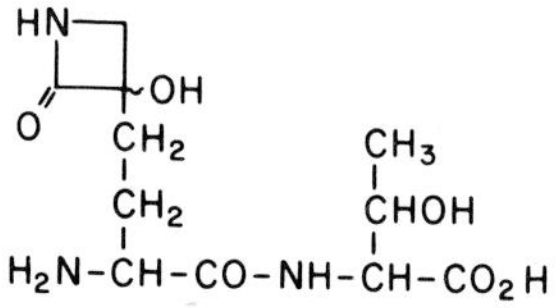 HN, O, OH CH_2 / CH_3 CH_2 / $CHOH$ $H_2N-CH-CO-NH-CH-CO_2H$	Stewart (1971) Taylor *et al.*, (1972)
[(2-serine)-tabtoxin]	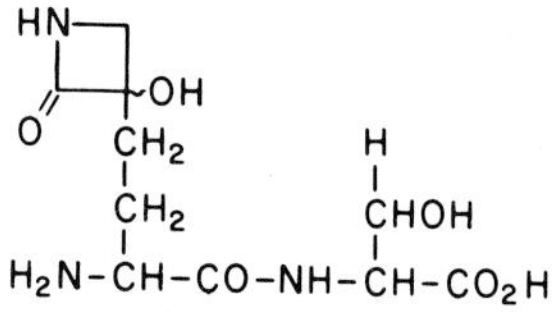 HN, O, OH CH_2 / H CH_2 / $CHOH$ $H_2N-CH-CO-NH-CH-CO_2H$	Taylor *et al.* (1972)
Rhizobium japonicum (rhizobitoxine)	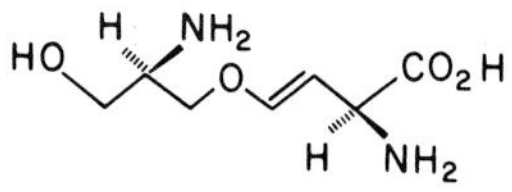 HO, H, NH_2, O, CO_2H, H, NH_2	Owens *et al.* (1972a,b), Keith (1975)
Pseudomonas syringae pv. *antropurpurea* (coronatine)	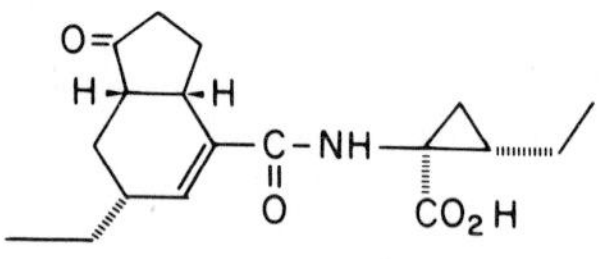 O=, H, H, C(=O)–NH, CO_2H	Ichihara *et al.* (1977), Ichihara *et al.* (1979)

TABLE 6. cont.

Organism (toxin)	Structure	Reference
Pseudomonas syringae pv. *tagetis* (tagetitoxin) p.v. manitrotis	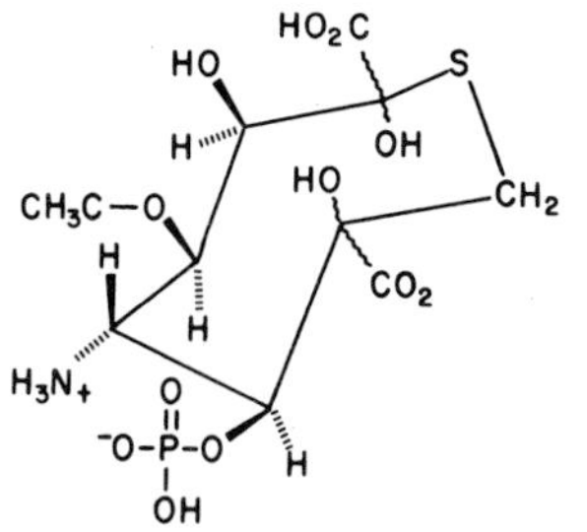	Mitchell and Hart (1983)
Pseudomonas syringae pv. *syringae* (syringomycin)	Peptide of unknown structure; ferric siderophore	Gross and DeVay (1977) Gross *et al.* (1977) Gross (1982)

V. Concluding Remarks

Toxins involved in plant disease include an interesting variety of molecular structures and perhaps a corresponding variety of physiological activities. Even the host-selective toxins, though biologically very narrowly defined, represent widely different chemical categories. Examples include peptides, glycosides, linear polyketols and esters (Table 1). In addition, most of the host-selective toxins occur in multiple molecular forms. Their non-toxic analogs and homologs, such as those described for HC- and HS-toxins, could serve as excellent controls in mode of action studies. These analogs or homologs are either naturally occurring, as in HS-toxin, and/or can be obtained by chemical modification of toxins or by chemical synthesis.

Toxins function as a chemical interface between pathogens and their host plants. Structural information is an important prerequisite for the study of the biosynthesis and the role of the toxins in pathogenesis. A knowledge of toxin identities will enable us to devise physico-chemical methods for monitoring their concentrations in parallel with bioassays. Laboratory synthesis, modification of the toxin, or introduction of radioactive label can facilitate studies on chemical transformation and mode of action of the toxin by very sensitive and quantitative means.

Firm evidence supporting several structural assignments has now been obtained. In at least one case, HC-toxin, elucidation of the structure has been independently accomplished in several laboratories, thus meeting the requirements suggested by Yoder (1981) for "...independent confirmation of reported observations, a fundamental element of experimental science...".

It can be expected that some of the toxins act at the lipid-water interfaces of membranes. Particularly good candidates are those compounds that are amphiphilic, having both polar and non-polar parts. The synthesis of models of these compounds should be possible as accomplished in other fields e.g. biologically active peptides (Kaiser and Kézdy, 1983).

Now that several important toxins, particularly host-selective toxins, have been structurally identified, research in the next few years should be especially productive in increasing our understanding of their biosynthesis and mode of action. The relative hiatus of the 1970s toxins research at a structural and molecular level appears to be overcome.

Acknowledgments

I wish to thank Prof. D. Arigoni and W. Acklin for help with the portion on *Helminthosporium sacchari* and numerous other suggestions. I also thank J.A.A. Renwick, O. Yoder, H. Mussell, R.C. Staples, R.P. Korf, and other colleagues for critically reading portions or all of the manuscript and Joanne Martin for expert secretarial aid. Research of the author was supported in part by grants from U.S. Department of Agriculture and NSF.

References

Alcorn, J.L. (1983). *Mycotaxon 17*, 1.
Anderegg, R.J., Biemann, K., Manmade, A., and Gosh, A.C. (1979). *Biomed. Mass Spectrom. 6*, 129.
Arigoni, D. (1975). *Pure and Appl. Chem. 41*, 219.

Ballio, A. (1981). *In* "Toxins in Plant Disease" (R.D. Durbin, ed.), p.395. Academic Press, New York.

Baltzer, P. and Johnston, R.B. (1982). Unpublished results.

Beechan, C.M., Djerassi, C., and Eggert, H. (1978). *Tetrahedron 34*, 2503.

Beier, R.C. (1980). Ph.D. Thesis, 334 pp. Montana State Univ.

Beier, R.C., Mundy, B.P., and Strobel, G.A. (1982). *Experientia 38*, 1312.

Bottini, A.T. and Gilchrist, D.G. (1981). *Tetrahedron Lett. 22*, 2719.

Bottini, A.T., Bowen, J.R., and Gilchrist, D.G. (1981). *Tetrahedron Lett. 22*, 2723.

Carr, S.A., Herlihy, W.C., and Biemann, K. (1981). *Biomed. Mass Spectrom. 8*, 51.

Ciuffetti, L.M., Pope, M.R., Dunkle, L.D., Daly, J.M., and Knoche, H.W. (1983). *Biochemistry 22*, in press.

Closse, A. and Huguenin, R. (1974). *Helv. Chim. Acta 57*, 533.

Comstock, J.C., Martinson, C.A., and Gegenbach, B.G. (1973). *Phytopathology 63*, 1357.

Daly, J.M. (1982). *In* "Plant Infection: Physiological and Biochemical Basis" (Y. Asada, W.R. Bushnell, S. Ouchi, and C.P. Vance, eds.) p.214. Japan Scientific Societies Press, Tokyo and Springer-Verlag, Berlin, Heidelberg, New York.

Daly, J.M. (1983). *In* "Proceedings of 5th International Congress of Pesticide Chemistry" (J. Miamoto, ed.) Pergammon Press, London (In press).

Daly, J.M. and Barna, B. (1980). *Plant Physiol. 66*, 580.

Daly, J.M. and Knoche, H.W. (1982). *In* "Advances in Plant Pathology", Vol. 1 (D.S. Ingram and P.H. Williams, eds.). Academic Press, New York.

Danko, S., Kono, Y., and associates. (1983). In preparation.

Dorn, F. and Arigoni, D. (1974). *Experientia 30*, 851.

Durbin, R.D. (ed.) (1981). "Toxins in Plant Disease". Academic Press, New York.

Durbin, R.D. (1982). *In* "Phytopathogenic Prokaryotes" (M.S. Mount and G.H. Lacy, eds.) Academic Press, New York.

Durbin, R.D. (1983). *In* "Biochemical Plant Pathology" (J.A. Callow, ed.), p.423. John Wiley and Sons, Inc., N.Y.

Duvick, J., Daly, J.M., Kratky, Z., Macko, V., Acklin, W. and Arigoni, D. (1983). *Phytopathology*, abstract submitted.

Gardner, J.M., Templeton, J.L., and King, R.W. (1982). *Phytopathology 72*, 973.

Gilchrist, D.G. and Grogan, R.G. (1976). *Phytopathology 66*, 165.

Godtfredsen, W.O. and Vandegal, S. (1964). *Proc. Chem. Soc. London*, 188.

Gross, D.C. (1982). *Phytopathology 72*, 941.

Gross, D.C. and DeVay, J.E. (1977). *Physiol Plant Pathol. 11*, 13.

Gross, D.C., DeVay, J.E. and Stadtman, F.H. (1977). *J. Appl. Bacteriol. 43*, 453.

Gross, M.L., McCrery, D., Crow, F., Tomer, K.B., Pope, M.R., Ciuffetti, L.M., Knoche, H.W., Daly, J.M., and Dunkle, L.D. (1982). *Tetrahedron Lett., 23*, 5381.

Hirota, A., Suzuki, A., Aizawa, K., and Tamura, S. (1973). *Agr. Biol. Chem. 37*, 955.

Ichihara, A., Shiraisi, K., Sato, H., Sakamura, S., Nishiyama, K., Sakai, R., Furusaki, A., and Matsumoto, T. (1977). *J. Am. Chem. Soc. 99*, 636.

Ichihara, A., Siraishi, K., Sakamura, S., Furusaki, A., Hashiba, N., and Matsumoto, T. (1979). *Tetrahedron Lett. 1979*, 365.

Ito, Y. (1981). *J. Biochem. Biophys. Methods 5*, 105.

Kaiser, E.T. and Kézdy, F.J. (1983). *Proc. Natl. Acad. Sci. U.S.A.*, in press.

Kanmera, T., Aoyagi, H., Waki, M., Kato, T., Izumiya, N., Noda, K., and Ueno, T. (1981). *Tetrahedron Lett. 22*, 3625.

Kawai, M., Jasensky, R.D. and Rich, D.H. (1983a). *J. Am. Chem. Soc.*, in press.

Kawai, M., Rich, D.H. and Walton, J.D. (1983b). *Biochem. Biophys. Res. Commun. 111*, 398.

Keith, D.D. (1975). *Tetrahedron 31*, 2629.

Kohmoto, K., Scheffer, R.P., and Whitesite, J.O. (1979). *Phytopathology 69*, 667.

Kono, Y. and Daly, J.M. (1979). *Bioorg. Chem. 8*, 391.

Kono, Y., Takeuchi, S., Kawarada, A., Daly, J.M., and Knoche, H.W. (1980a). *Agric. Biol. Chem. 44*, 2613.

Kono, Y., Takeuchi, S., Kawarada, A., Daly, J.M., and Knoche, H.W. (1980b). *Tetrahedron Lett. 21*, 1537.

Kono, Y., Knoche, H.W., and Daly, J.M. (1981a). *In* "Toxins in Plant Disease" (R.D. Durbin, ed.), p.109. Academic Press, New York.

Kono, Y., Takeuchi, S., Kawarada, A., Daly, J.M., and Knoche, H.W. (1981b). *Agric. Biol. Chem. 45*, 2111.
Kono, Y., Takeuchi, S., Kawarada, A., Daly, J.M., and Knoche, H.W. (1981c). *Bioorg. Chem. 10*, 206.
Kupchan, S.M., Streelman, D.R., Jarvis, B.B., Dailey, R.G., Jr. and Sneden, A.T. (1977). *J. Org. Chem. 42*, 4221.
Lee, S., Aoyagi, H., Shimohigashi, Y., Izumiya, N., Ueno, T., and Fukami, H. (1976). *Tetrahedron Lett. 843*.
Liesch, J.M., Sweeley, C.C., Staffeld, C.D., Anderson, M.S., Weber, D.J. and Scheffer, R.P. (1982). *Tetrahedron 38*, 45.
Livingston, R.S., and Scheffer, R.P. (1981a). *Phytopathology 71*, 891.
Livingston, R.S., and Scheffer, R.P. (1981b). *J. Biol. Chem. 256*, 1705.
Livingston, R.S., and Scheffer, R.P. (1982). *Phytopathology 72*, 933.
Livingston, R.S., and Scheffer, R.P. (1983). *Plant Physiol.*, in press.
Macko, V., Arigoni, D., and Acklin, W. (1981a). *Phytopathology 71*, 892.
Macko, V., Goodfriend, K., Wachs, T., Renwick, J.A.A., Acklin, W. and Arigoni, D. (1981b). *Experientia 37*, 923.
Macko, V., Acklin, W., and Arigoni, D. (1982a). *American Chemical Society, 183rd Natl. Meeting, Abstr. 252.*
Macko, V., Grinnalds, C., Golay, J., Arigoni, D., Acklin, W., Weibel, F., and Hildenbrand, C. (1982b). *Phytopathology 72*, 942.
Macko, V., Acklin, W., Hildenbrand, C., Weibel, F., and Arigoni, D. (1983a). *Experientia 39*, 343.
Macko, V., Arigoni, D., and associates. (1983b). *In* preparation.
Mascagni, P., Pope, M.R., Ciufetti, L.M., Knoche, H.W., and Gibbons, W.A. (1983). *Biochem. Biophys. Res. Commun. 112*, in press.
Mitchell, R.E. (1976a). *Nature 260*, 75.
Mitchell, R.E. (1976b). *Phytochemistry 15*, 1941.
Mitchell, R.E. (1981). *In* "Toxins in Plant Disease" (R.D. Durbin, ed.) p.259. Academic Press, New York.
Mitchell, R.E. and Durbin, R.D. (1981). *Physiol. Plant Pathol. 18*, 157.
Mitchell, R.E. and Hart, P.A. (1983). *Phytochemistry, 22*, in press.
Mitchell, R.E. and Parsons, E.A. (1977). *Phytochemistry 16*, 280.

Nakashima, T., Ueno, T., and Fukami, H. (1982). *Tetrahedron Lett. 23*, 4469.
Novacky, A. and Ullrich-Eberius, C.I. (1982). *Physiol. Plant Pathol. 21*, 237.
Nishimura, S. and Kohmoto, K. (1983). *Annu. Rev. Phytopathol. 21*, in press.
Okuno, T., Ishita, Y., Nakayama, S., Sawai, K., Fumita, T., and Sawamura, K. (1974a). *Ann. Phytopathol. Soc. Japan 40*, 375.
Okuno, T., Ishita, Y., Sawai, K., and Matsumoto, T. (1974b). *Chem. Lett.* 635.
Okuno, T., Ishita, Y., Sugawara, A., Mori, Y., Sawai, K. and Matsumoto, T. (1975). *Tetrahedron Lett.* 335.
Owens, L.D., Thompsons, J.F. and Rennessey, P.V. (1972a). *J. Chem. Soc. Chem. Commun. 1972*, 715.
Owens, L.D., Thompson, J.F., Pitcher, R.G., and Williams, T. (1972b). *J. Chem. Soc. Chem. Commun. 1972*, 714.
Pastuszak, J., Gardner, J.H., Singh, J.S. and Rich, D.H. (1982). *J. Org. Chem. 47*, 2982.
Patil, S.S., Kwok, O.C.H. and Moore, R.E. (1983). *Toxicon.* In press.
Pope, M.R., Ciuffetti, L.M., Knoche, H.W., McCrery, D., Daly, J.M., and Dunkle, L.D. (1983a). *Biochemistry*, in press.
Pope, M.R., Mascagni, P., Ciuffetti, L.M., Knoche, H.W., and Gibbons, W.A. (1983b). Submitted for publication.
Pringle, R.B. (1970). *Plant Physiol. 46*, 45.
Pringle, R.B. (1971). *Plant Physiol. 48*, 756.
Pringle, R.B. and Scheffer, R.P. (1967). *Phytopathology 57*, 1169.
Rich, D.H. (1981). *In* "Toxins in Plant Disease" (R.D. Durbin, ed.), p.295. Academic Press, New York.
Rich, D.H., Kawai, M. and Jarenski, R.D. (1983). *Int. J. Pep. Protein Res.*, in press.
Rietschel-Berst, M., Jentoft, N.H., Rick, P.D., Pletcher, C., Fang, F. and Gander, J.E. (1977). *J. Biol. Chem. 252*, 3219.
Rudolph, K. (1976). *In* "Physiological Plant Pathology" (R. Heitefuss and P.H. Williams, eds.), p.270. Springer-Verlag, Berlin and New York.
Scheffer, R.P. (1976). *In* "Physiological Plant Pathology (R. Heitefuss and P.H. Williams, eds.), p.247. Springer-Verlag, Berlin.
Scheffer, R.P. and Livingston, R.S. (1980). *Phytopathology 70*, 400.
Schröter, H., Macko, V., Acklin, W. and Arigoni, D. (1982). Unpublished results.

Schröter, H., Novacky, A. and Macko, V. (1983). *Plant Physiol. Suppl.*, in press.
Shimohigashi, Y., Lee, S., Kato, T., Izumiya, N., Ueno, T., and Fukami, H. (1977). *Agric. Biol. Chem. 41*, 1533.
Silver, D.J. and Gilchrist, D.G. (1982). *J. Chromatog. 238*, 167.
Smaley, E.B. and Strong, F.M. (1974). *In* "Mycotoxins" (I.F.H. Purchase, ed.), p.199. Elsevier, Amsterdam.
Stähelin, H. and Trippmacher, A. (1974). *J. Cancer 10*, 801.
Steiner, G.W. and Byther, R.S. (1971). *Phytopathology 61*, 691.
Steiner, G.W. and Strobel, G.A. (1971). *J. Biol. Chem. 246*, 4350.
Stewart, W.W. (1971). *Nature 229*, 174.
Stoessl, A. (1981). *In* "Toxins in Plant Disease" (R.D. Durbin, ed.), p.109. Academic Press, New York.
Strobel, G.A. (1982). *Annu. Rev. Biochem. 51*, 309.
Suzuki, Y., Knoche, H.W. and Daly, J.M. (1982a). *Bioorg. Chem. 11*, 300.
Suzuki, Y., Tegtmeier, K.J., Daly, J.M. and Knoche, H.W. (1982b). *Bioorg. Chem. 11*, 313.
Suzuki, Y., Danko, S.J., Daly, J.M. and Knoche, H.W. (1983). In preparation.
Tamm, C.H. and Breitenstein, W. (1980). *In* "The Biosynthesis of Mycotoxins" (P.S. Steyn, ed.). p.69. Academic Press, New York.
Taylor, P.A., Schnoes, H.K. and Durbin, R.D. (1972). *Biochim. Biophys. Acta 286*, 107.
Tegtmeier, K.J., Daly, J.M., and Yoder, O.C. (1982). *Phytopathology 72*, 1492.
Tipton, C.L., Paulsen, P.V., and Betts, R.E. (1977). *Plant Physiol. 59*, 907.
Uchytil, T.F. and Durbin, R.D. (1980). *Experientia 36*, 301.
Ueno, T., Hayashi, Y., Nakashima, T., Fukami, H., Nishimura, S., Kohmoto, K. and Sekiguchi, A. (1975a). *Phytopathology 65*, 82.
Ueno, T., Nakashima, T., Hayashi, Y., and Fukami, H. (1975b). *Agric. Biol. Chem. 39*, 1115.
Ueno, T., Nakashima, T., Hayashi, Y., and Fukami, H. (1975c). *Agric. Biol. Chem. 39*, 2081.
Ueno, T., Nakashima, T., and Fukami, H. (1982). *In* "Plant Infection: The Physiological and Biochemical Basis" (Y. Asada, W.R. Bushnell, S. Ouchi, and C.P. Vance, eds.) p.235. Japan Sci. Soc. Press, Tokyo/Springer-Verlag, Berlin.

Ueno, T., Nakashima, T., and Fukami, H. (1983). *In* "Proceedings of 5th International Congress of Pesticide Chemistry, 1982" (J. Miyamoto, ed.), Pergamon Press, London. (In press).

Walton, J.D. and Earle, E.D. (1983). *Physiol. Plant Pathol.*, in press.

Walton, J.D., Earle, E.D., and Gibson, B.W. (1982). *Biochem. Biophys. Res. Commun. 107*, 785.

Weibel, F., Acklin, W. and Arigoni, D. (1982). Unpublished results.

Wheeler, H. and Luke, H.H. (1963). *Annu. Rev. Phytopathol. 17*, 223.

Wolpert, T.J. and Dunkle, L.D. (1980). *Phytopathology 70*, 872.

Yoder, O.C. (1973). *Phytopathology 63*, 1361.

Yoder, O.C. (1980). *Annu. Rev. Phytopathol. 18*, 103.

Yoder, O.C. (1981). *In* "Toxins in Plant Disease" (R.D. Durbin, ed.), p.45. Academic Press, New York.

3

Molecular Modes of Action[1]

D. G. GILCHRIST

I. Introduction

It is generally accepted that individual plants encounter disease stress through a series of steps beginning with the arrival of inoculum at the plant surface, proceeding through a range of physiological impacts, and ending with macroscopic symptoms characteristic of a given disease. The scope of attacking mechanisms used by pathogens to effect physiological damage include macromolecules involved in vascular plugging, enzymes involved in the digestion of polymeric host components, and metabolites which may be toxic to host cells (pathotoxins) or act as phytohormones.

Studies on the mode of action of pathotoxins historically have been accorded great promise for yielding details of the chemical mechanisms of pathogen induced stress at the level of plant cell metabolism. In fact, presumed and documented involvement of pathotoxins in plant diseases challenged the imagination and investigative efforts of plant pathologists even before Gauman first proposed that all plant diseases ultimately derive from a toxicologic basis (Gauman, 1954). Apparently the only unifying impact of this proposal in its time was the near universal skepticism exhibited by the author's peers. At the other extreme is the

[1]*Some unpublished data presented in this article were obtained through funds from Grant No. 59-2063-01406 CRGO, SEA, U.S. Department of Agriculture and NSF Grant No. PCM 80-1173.*

TOXINS AND PLANT PATHOGENESIS
ISBN 0 12 200780 8

sentiment of Day who suggested that pathotoxins are unique biological accidents of little overall significance (Day, 1979). However, time and effort have produced a growing catalog of accepted examples in support of Gauman's thesis but have not provided the preponderance of evidence necessary to disprove the negative hypothesis of Day.

Pathotoxins have been categorized historically at several levels including their organismal origin, the similarity and differences of their chemical structures, the symptoms they elicit, their apparent role in disease as primary or secondary determinants, and their relative role as virulence factors (Scheffer and Briggs, 1981). Unfortunately, none of the similarities or differences in the above categories have provided reliable clues to molecular modes of action. In fact, common physiological effects such as ion leakage may have produced unwarranted optimism about the widespread role of plasma membranes as a site of action for the special class of pathotoxins referred to as host-specific toxins (Wheeler, 1982).

The ultimate objective of studies of pathotoxin action is not restricted to characterization of the specific molecular interaction between toxins and host targets in chemically specific terms. It also requires the complete description of the sequence (and direction) of the array of perturbed metabolic and physiological symptoms characteristic of disease which arise from the primary interactions. In view of this, pathotoxins have been chosen to study the sequence of physiological events in pathogenesis since they: 1) avoid alterations due to the actual growth of the pathogen in diseased tissue subsequent to the determining event; 2) permit the reduction in time of the stress determining events from days to hours when the requirement for pathogen ingress and metabolism is circumvented; 3) allow experimental use of tissues (host and non-host) which normally are not colonized by the respective pathogens.

In spite of these attributes, unambiguous assignment of either descriptive roles or specific chemical sites of involvement in disease syndromes have been resolved for only a few pathotoxins. The fact that they appear to represent direct chemical probes is perhaps part of the problem. Several difficulties limiting precise conclusions are now apparent from current literature: 1) as a group pathotoxins are incompletely characterized and structurally diverse where characterized, 2) the ability to obtain chemically pure preparations is not inherently simple, 3) the assays available usually do not preclude numerous possible secondary events nor multiple targets within the plants, and 4) genetic control of differential host reaction is, in most cases, insufficiently described to avoid "pleotrophic" confounding due to the use of genotypically heterogenous tissue.

II. Scope and Organization

This review deals with methods and evidence leading to the assignment of molecular modes of pathotoxin action where the proposed site of direct chemical interaction in the host is consistent biologically, physiologically, and genetically with the disease symptoms expressed. In this context, pathotoxins are defined as chemical products of pathogen metabolism which are recognized by their ability to reveal one or more consequence of metabolic dysfunction in susceptible host tissue. Other chapters will assess the definitional and physiological limits placed on particular pathotoxins and their role(s) in disease. In addition, the literature on the role(s) of toxins in plant disease has been covered quite thoroughly in several recent review articles (Yoder, 1980; Daly and Knoche, 1982), books (Durbin, 1981), and published symposia (Daly and Uritani, 1979) on the topic.

Past literature and recent reviews have emphasized the role of membrane dysfunction as an early and presumed primary basis for toxin action. This approach has not been successful to date, possibly because account is not taken of events occurring between the primary target and the symptoms ultimately expressed (such as ion leakage). It appears that two general ways pathotoxins may act are by combining with a primary target to directly induce physiological stress or by combining with a primary target which subtly influences host metabolism through a series of steps to provide an equally dramatic ultimate impact.

Review of recent literature on this topic suggests that the latter effect has not been considered widely. In constrast, the literature on mechanisms regulating intermediary metabolism of plants has expanded greatly in the last few years. A number of key regulatory mechanisms have been described which may be relevent to molecular modes of toxin action, especially in situations where membrane dysfunction may be a secondary consequence of the primary toxin interaction.

In view of this, the discussion in this chapter will take a slightly different tack than other recent reviews. With the aid of selected examples, an attempt will be made to illustrate experimental approaches used by several workers to resolve primary target sites of pathotoxins affecting critical steps in cellular metabolism. The discussion of specific examples will be preceded by a description of several key mechanisms reported to regulate intermediary metabolism in plants which could lead to grave metabolic consequences if perturbed by a pathotoxin.

Assignment of site-specific roles in disease to pathotoxins presupposes several experimental requirements before such studies can be unambiguously interpreted. Hence, a number of highly desirable experimental controls or prerequisites will be discussed. Some of the inherent pitfalls and limitations therein have been reviewed in detail by others (Daly, 1981; Daly and Knoche, 1982; Mitchell, 1981; and Yoder, 1980 and 1981) and will be summarized only briefly. This discussion is centered on potential sites of toxin effects within the host and will focus on specialized regulatory functions within the cell which have potential cellular impacts beyond the target site.

III. Intermediary Metabolism and Its Regulation

A. *General Concepts and Applications*

1. The Metabolic Steady State. The diversity and complexity of cells of living organisms requires metabolic regulation at several physical and chemical levels, any or all of which may be relevant to the perturbed metabolism associated with a host-parasite interaction. It follows that any unrelieved disruption of the steady-state between the vast network of energy-generating catabolic pathways and the myriad of energy-demanding biosynthetic reactions in the living cell, regardless of the cause, results in a "diseased state".

Intuition suggests host metabolic regulation will be altered first at the molecular site of action in host-toxin interactions. The objective, of course, is to learn precisely where a pathotoxin interacts with the host cell in a chemically definable fashion and what amplified sequence of secondary metabolic events leads to the characteristic morphological or physiological disease symptoms. These symptoms may include wilting, chlorosis, necrosis, or increases in ion leakage and in respiration. However, as noted earlier, experience has shown that deduction of the host site or sites of pathotoxin interaction from the physical or physiological symptoms is not easily achieved.

Several authors recently have detailed the limited value of symptoms such as wilting, chlorosis, necrosis, and growth distortions in predicting molecular mechanisms of action (Daly, 1981; Durbin, 1982). It would seem that there are two general factors which should be considered when attempting to use symptoms as molecular site predictors: 1) the type of molecular events known to impinge on the observed symptoms from the plant biochemical literature and 2) the time-course relationship expressed between toxin treatment of the plant tissue and appearance of various symptoms. In the case of the former, for example, adequate literature on the biochemical regulation of stomatal aperture exists from which testable experimental hypothesis on toxin induced wilting can be developed (Rashke, 1975).

Clues to the action of fusicoccin (FC) came from the ability of FC to regulate the stomatal apparatus (Turner and Grantiti, 1969). This observation complimented later work (Turner, 1972, 1973) indicating that within 1 hr FC altered K^+ in flux in leaves mimicking the flux of K^+ observed during normal stomatal regulation (Fisher, 1968). Further, FC mediated K^+ changes in corn root tissues (Marre, 1979). These results and others discussed by Daly (Daly, 1981) have now focused attention on K^+ activated ATPase from isolated membrane fractions which was activated 20-35% by administered FC (Beffagna et al., 1977). The precise molecular action of FC still is not completely resolved but this work illustrates a synthetic approach starting from a wilt symptom which then illuminated promising directions for future research. The existence of a chemically characterized structure of FC (Ballio et al, 1964) for use in these studies should be underscored.

The utility of chlorosis and necrosis on the other hand has been less applicable, in part, because these symptoms usually develop several days after cellular contact with purported toxins. The frequent observation that up to 24 hr is required before symptoms appear suggests that these symptoms are more likely the complex secondary consequence of some undisclosed primary interaction. The revealing studies with tentoxin, the chlorosis inducing toxin from *Alternaria alternata*, to be discussed later in relation to a proposed CF_1-ATPase site of action for this toxin should temper somewhat a completely pessimistic outlook on developing an experimental strategy from symptoms such as chlorosis. Similarly, growth distortions contain target-site clues which are, at this point, well hidden but which conceivably could lead to direct testing of specific target sites if sufficient literature existed on parallel actions of known compounds (such as hormones) in standardized systems.

Biochemical or physiological symptoms also have proven generally inconclusive in targeting sites or mechanisms responsible for the altered metabolic steady-states in pathotoxin perturbed cells. It seems logical to assume that changes in parameters such as endogenous ion leakage or respiration, especially if they occur shortly after toxin administration, would prove useful. However, extensive results derived from assays of ion leakage have tended to focus attention on possible alterations in plasmalemma permeability without, as yet, providing definitive conclusions about either a target site or a role for altered permeability as the primary mechanism of cell stress. This is notwithstanding the fact that alterations in endogenous leakage have been useful as assay procedures for several toxins including victorin, PC-toxins, and T-toxins (Yoder, 1981). These and other physiological considerations have been tantalizingly suggestive but fruitless to date as probes of molecular targets for both fungal and bacterial toxins.

These brief reflections on situations where readily available preliminary clues have given unclear results at either the morphological or biochemical level justify consideration of some of the problems associated with attempting to probe the plant's metabolic maze. Conducting specific molecular-target studies with toxin preparations of marginal purity simply adds to the confusion.

2. *Regulation of the Metabolic Steady-State*. It seems appropriate to discuss briefly some of the known mechanisms and sites of regulatory control from which pathotoxin impacts might emanate as a prelude to consideration of sites which have been proposed. In the past two decades a marked increase in information has appeared on enzymatic mechanisms regulating carbon flux through individual anabolic and catabolic pathways (usually observed by *in vitro* kinetic experiments) along with principles of coordinate regulation by cross-pathway interlocks. However, the possible impact of their alteration in response to infection (or pathotoxins) is less clearly characterized. Kosuge and Kimple (1982) recently reviewed much of the literature on altered plant metabolism in response to disease.

The existing literature on the factors and mechanisms involved in maintenance of the metabolic state reveals an awesome array of potential target sites where a pathotoxin could easily perturb metabolism of the plant cell. It is important to anticipate that even subtle site-specific toxin action can be readily amplified over this same vast network of host metabolic pathways. Without resorting to unlimited speculation, the final sequence and consequence of such secondary amplification of toxin action cannot be described in any specific case at present. In cases to be discussed later where site specific toxin activities have been reported, testable metabolic consequences contributing to the observable symptoms can be constructed (Daly, 1981; Durbin, 1981; Daly and Knoche, 1982; Kosuge and Kimple, 1982). The testing of such hypotheses is critical for assignment of a relative role to the proposed site, at least as far as toxin induced stress is concerned.

a. Regulation of enzyme activity. The direct inhibition of a key enzyme(s) of primary metabolism is, perhaps, the simplest target site to visualize for a purported pathotoxin. As noted by Durbin (Durbin, 1981) it has been stated that "for every enzymatic reaction the biochemist wishes to study, there is, in nature, a compound uniquely designed to serve as its specific inhibitor" (Lardy, 1980). It is worth noting, although by no means a unifying dogma, that all of the toxin-specific sites for which solid evidence presently exists involve inhibition of a different, specific, and metabolically important enzyme.

The subtlty and/or kinetic complexity with which inhibition of enzyme activity in plants can occur is not only of direct biological importance to the plant but, from an experimental standpoint, an indirect potential limitation to detection of a primary toxin-enzyme interaction. In general terms there are at least five possible means by which enzyme mediated carbon flux through a given pathway can be regulated: 1) the supply of an initial substrate, 2) commitment of an initial substrate to further metabolism through subsequent pathways, 3) supply of energy, coenzymes, or cofactors in coupled reactions, 4) the absolute concentration of an enzyme, and 5) the relative activity of an enzyme under modulation by allosteric mechanisms involving products of either branched or interlocked pathways. Thus, it follows that any situation which alters enzyme velocity of a particular reaction could serve to perturb the entire integrated system of metabolism (Kosuge and Gilchrist, 1976).

Analysis of individual enzyme-catalyzed steps in metabolism reveals two general catagories: 1) single-sited enzymes whose activity operates in the cell at or near thermodynamic equilibrium in the presence of both substrates and products and 2) enzymes which, for various kinetic reasons, do not establish equilibrium readily and possess one or several regulatory site(s) able to bind reactive metabolites (ligands) other than the substrate (Davis, 1980).

In the case of enzymes operating at equilibrium, simple binding of the substrate on a one-sited enzyme constitutes a primitive regulatory mechanism. Under equilibrium conditions the binding function is a rectangular hyperbola (Fig. 1) wherein the steepness of the curve decreases when the ligand concentration increases (Segel, 1975). Hence, at substrate concentrations ([S]) below the dissociation constant (K_d) of the enzyme-substrate complex a small increase in substrate concentration will produce a large increase in binding versus the situation above K_d where large increases in [S] produce only moderate increases in binding and the response is insensitive to changes in substrate concentration. This leads to the classical Michaelis-Menten law where the enzyme responds sharply to substrate concentrations below its Km but less so above the Km. Because it is believed that intracellular substrate concentrations normally exist near the Km of such enzymes (Cleland, 1967), this type of control serves to maintain an even flux through sequential enzyme catalyzed steps functioning near thermodynamic equilibrium in the cell. Other factors which effect the rate of such enzymes include, besides substrate and product concentrations, the availability of energy in the form of ATP (or ADP) and cofactors such as pyridine nucleotides NADH or NADPH.

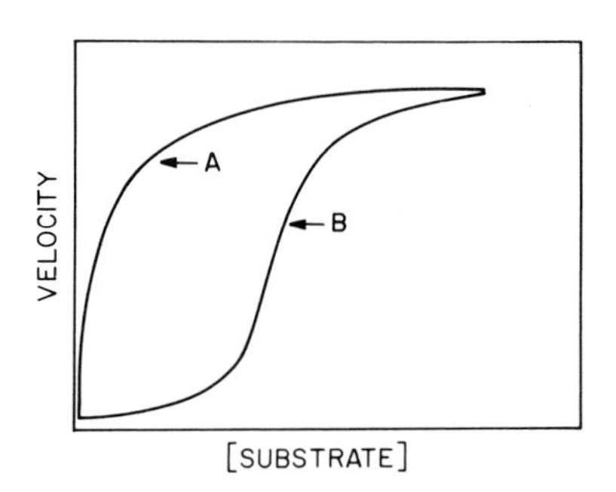

FIGURE 1. The relationship between enzyme activity and substrate concentration for two general classes of enzymes. Curve A represents a simple one-sited (equilibrium) or non-regulatory enzyme. Curve B represents a multi-sited (non-equilibrium) or regulatory enzyme. Km (see text) equals the substrate concentration at one-half maximum velocity for Curve A.

For pathotoxins to directly affect the activity of such enzymes they would seemingly have to compete with the substrate at the active site (either competitively or noncompetitively). This would require toxins with structural configurations similar to the substrates or reaction products. Alternatively, and indirectly, the action of toxins could alter the availability of the substrate, cofactors, or the rate of enzyme turnover. This latter effect would imply that the toxin-specific interaction is at a preceding enzyme.

Binding of ligands to the second (non-equilibrium) class of enzymes presents a markedly different format for regulation. There is almost no response of the enzyme to changes in substrate concentration below a certain value but a maximum response is obtained at substrate concentrations above that value (Fig. 1). It is generally accepted that enzymes at non-equilibrium steps play a predominant role in overall control of pathway rates because the potential activity of "equilibrium enzymes" is far in excess of the normal steady state flux. The study of mechanisms influencing enzyme activity is, therefore, concentrated on the non-equilibrium or "regulatory" enzymes. Factors influencing the activity of regulatory enzymes include the presence of positive and negative effectors specific for given enzymes in addition to the availability of substrates, cofactors, energy status and pH of the cellular environment.

Cursory examination of pathway relationships common to both primary and secondary plant metabolism reveals the existence of numerous multibranched and interconnected pathways. Regulatory enzymes appear to be situated in key locations to control and respond to flux through this complex network (Ricard, 1980). They are generally found to catalyze the first committed step in a particular pathway or branch point and hence, through regulation of product formation at that step, control substrate availability to succeeding (equilibrium) steps. This latter function is served by feedback inhibition by one or more end products of the pathway involved (monovalent feedback). In addition, response to end products of different but related pathways (polyvalent feedback) determines a level of integrated regulation by cross-pathway metabolites, a function known as metabolic interlock (Jensen, 1969).

Kinetic responses to effectors can be either positive or negative. One of the most common positive effectors of regulatory enzymes is the substrate itself which operationally serves to increase enzyme activity as the substrate concentration increases, a relationship which relates to the concept of allostery.

Allosteric behavior toward the substrate is revealed kinetically when a plot of reaction velocity vs substrate concentration ([S]) is sigmoidal rather than the rectangular hyperbola noted earlier for nonregulatory enzymes (Segal, 1975). The sigmoid curve (Fig. 1) has been interpreted to indicate the presence of multiple substrate binding sites which interact cooperatively. The binding of one substrate molecule facilitates the binding of a second and so on. The physiological significance is that there is an threshold substrate concentration below which the enzyme is insensitive to large changes in [S] but above which small changes in [S] produce relatively large changes in activity. Such curves have been observed for both allosteric activators and inhibitors (Segel, 1975).

The general concept of allostery also emphasizes the fact that most metabolic activators and inhibitors bear little structural resemblance to the substrate at the first committed step in a pathway. This, plus the knowledge that most regulatory enzymes are composed of more than one subunit (and may even exist in multifunctional enzyme aggregates), led to the postulate that metabolic effectors (positive and negative) react at a site (effector site) on the enzyme different from the substrate site. In the case of an inhibitor, binding at the effector site induces or stabilizes a conformational change in the enzyme which results in a lower affinity or absolute activity. The opposite effect would occur for any activator of an allosteric enzyme, including its substrate.

The existence of multiple substrate and effector sites on a regulatory enzyme affords a very sensitive and subtle regulatory format. It also affords an equally subtle and potentially significant mechanism for nonsubstrate-site interaction by pathotoxins with little or no resemblance to the substrate, but with a structural relationship to the effectors.

In addition to the conceptually logical effector relationships described above there are situations where sites with no apparent relationship to either substrates or products exist on the enzyme and which affect the allosteric behavior of enzymes. For example, Yon has reported a hydrophobic binding site on aspartate carbamoyltrasferase from wheat germ where binding of compounds such as deoxycholate induces an allosteric transition state of lowered enzyme activity (Yon, 1973). Existence of such evolutionarily conserved, but metabolically mysterious, sites affords yet another potential site for toxin interaction. Unfortunately, the existence of the latter category of sites, if widespread, presents a deductive dilemma for experimentally locating such sites from physiological symptoms.

Even if one focuses on pathotoxins of known structure and attempts to analyze potential effects on regulatory enzymes where structural complimentary is suspected at either the substrate or effector sites, there are additional difficulties to consider. As was indicated earlier, the substrate and allosteric effector molecules occupy distinct sites in the enzyme and, in fact, the distinctive regulatory properties of the enzyme may be altered independently of the catalytic activity (Mankovitz and Segal, 1969). Some factors which can cause differential desensitization of regulatory enzymes include temperature, pH, ionic strength, protein concentration, as well as dialysis, treatment with arsenicals, urea, and proteolytic enzymes (Cohen, 1968). Loss of regulatory properties also has been reported following the use of acetone powders during the extraction of regulatory enzymes from plant tissues (Gilchrist et al, 1972).

Conversely, it is important to appreciate the fact that the presence of regulatory properties, including the sigmoid substrate saturation curve, may be effectively masked by the presence of effector molecules with high-binding constants. The age of the tissue source can drastically affect the regulatory properties of an enzyme _in vitro_ as well as _in vivo_ as was demonstrated in the case of phosphofructokinase from developing Brussels sprouts (Dennis and Coultate, 1967). The enzyme from the most immature tissues showed the greatest sensitivity to inhibition by ATP (50% inhibition at 2 mM) while that from mature and senescent leaves required at least 4-fold higher concentration of ATP to provide equal inhibition. Thus, continual monitoring for loss of critical _in vivo_ regulatory properties along with catalytic activity is mandatory in all regulatory enzyme studies.

Another factor complicating the study of possible toxin induced alteration in enzyme function is the occurrence of multiple forms of an enzyme, each with different regulatory properties. For instance, etiolated mung bean seedlings were found to possess two separable forms of chorismate mutase which were further distinguished by their response to several allosteric effectors (Gilchrist et al., 1972). Chorismic acid, a branch point intermediate in the shikimic acid pathway, is a precursor for separate pathways committed to either the synthesis of the aromatic amino acids phenylalanine and tyrosine by chorismate mutase or the synthesis of tryptophan by anthranilate synthetase (Gilchrist and Kosuge, 1980). Regulation of chorismate metabolism through

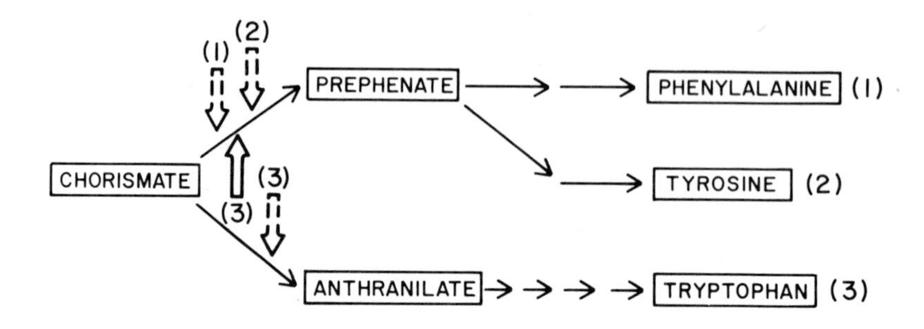

FIGURE 2. Regulation of chorismate metabolism in plants. Phenylalanine (1) and tyrosine (2) feedback inhibit (dashed arrows) chorismate mutase. Tryptophan (3) feedback inhibits (dashed arrow) anthranilate synthetase and activates (solid arrow) chorismate mutase as well as antagonizing the inhibition of chorismate mutase by phenylalanine and tyrosine thus providing interlocked control of this branched pathway.

these competing pathways involves both feedback inhibition and activation (Fig. 2).

In the simpler case, anthranilate synthetase is feedback inhibited by tryptophan, the end product of the pathway initiated by this enzyme. Regulation at the chorismate mutase step, on the other hand, is more complicated. One of the chorismate mutase isoenzymes (CM-1) from mung bean appeared to be directly involved in regulating the flow of chorismate through the respective branch pathways (Gilchrist and Kosuge, 1974). It displayed sigmoid substrate saturation kinetics (activation), was specifically feedback-inhibited by phenylalanine and tyrosine, and was activated by tryptophan, a cross-pathway metabolite. Furthermore, tryptophan antagonized the feedback inhibition by either phenylalanine or tyrosine. The other form of the enzyme (CM-2) represented 40% of the total extractable chorismate mutase activity, but was completely insensitive to feedback regulation by any of the three aromatic amino acids (Gilchrist and Kosuge, 1975). While the physiological role of the CM-2 form is not clear, recognition of its presence and separation of the two forms would be crucial in *in vivo* or *in vitro* studies where potential toxin mediated changes might be evaluated.

b. Regulation of enzyme concentration. The concentration of an enzyme in the cell at any given time is a function of the relative rates of the synthesis/degradation or activation/inactivation of the enzyme. In contrast to bacteria, studies with animals (Shimke, 1969) and plants (Zeilke and Filner, 1971) reveal that differential rates of enzyme synthesis and degradation are brought about by several factors including abrupt environmental changes (Filner, 1971), administration of hormones (Briggs, 1973), and developmental changes such as germination (Turner and Turner, 1975).

Control of apparent enzyme activity by changes in enzyme concentration is usually referred to as "coarse" control in contrast to the "fine" control afforded by regulatory enzymes responding to subtle changes in concentrations of metabolic intermediates (Turner and Turner, 1980). For example, amylases in some plant tissues appear to be under course control. α- and β-amylases in cereals both increased during germination (Shinke and Mugibayashi, 1972). Increases in α-amylase activity were due to de novo synthesis while β-amylase increases were due to activation of a constuitive inactive zymogen form. However, most of the total amylase activity during germination was due to the de novo synthesis of α-amylase (Shinke and Mugibayashi, 1972). Activity of starch phosphorylase, another glycolytic enzyme involved in starch degradation, also is subject to coarse control which is thought to be the predominant mechanism controlling its activity (Turner and Turner, 1980).

Mechanisms involving apparent induction or repression of enzyme synthesis conceivably could play a role in toxin mediated changes in enzyme activity, although the response time would be longer than those due to direct toxin binding to a constitutive enzyme. However, such mechanisms providing for changes in enzyme concentration could be just as chemically specific. Recent reports by Key et al. (1981) describe unique "heat shock" mRNA and protein synthesis in soybean cotyledons following a sudden exposure to elevated temperatures. It is not unrealistic to imagine stress response caused by foreign "chemicals" also inducing unique transcriptional products.

It may be significant that purified PC-toxin has recently been observed to ellicit increased specific mRNA and protein synthesis within 2 hr after toxin administration in susceptible but not resistant near-isogenic sorghum clones (L. D. Dunkle personal communication). Regardless, alteration in either transcription or translation processes by pathotoxins exists as a possible mechanism for ultimately perturbing the metabolic steady-state. As pointed out earlier by Daly (1980) selective interference with the synthesis of specific enzymes cannot be ruled out as a possible molecular mode of toxin action.

c. Regulation by pH. The relationship between cellular pH and in vitro enzyme activity represents a biological dilemma. It appears necessary that cellular pH be regulated within a limited range if the in vitro curves relating enzyme activity to pH have any in vivo significance. The general in vitro response of enzyme activity to pH, canonized in introductory biochemistry texts as a bell shaped curve, indicates that at pH values near its optimum an enzyme is relatively insensitive to small pH changes (± 1 pH unit). However, on either side of this limited optimum pH range enzyme activity decreases rapidly. Individual enzyme pH optimum range from at least pH 6 to pH 10, a range which would require numerous pH microclimates in the cell (or organelle) if all enzymes functioned at their optimum pH. Particularly troublesome is the fact that even enzymes in the same pathway do not always have complementary pH optima. Hence, many enzymes function at pH's away from their pH optimum where small changes in pH would be expected to cause large changes in potential activity.

Davies (1979, 1980) has stressed the role of metabolic pH-stats in providing a buffered cytoplasmic pH environment. The dual presence of malic enzyme and phosphenolpyruvate carboxylase (PEP carboxylase) in cytoplasm is an example of a metabolic pH-stat. Malic enzyme has an acid pH optima (6.5) whereas PEP carboxylase has an alkaline pH optima (7.5). If both enzymes are present in the same cellular vicinity they can function to maintain a stable pH somewhere between their respective optima. For instance, if the pH rises the carboxylating reaction would increase, converting oxalacetate to malate via PEP carboxylase. The conversion of malate to pyruvate (malic enzyme) would will decrease, producing a net gain of carboxyl groups, thus buffering the pH downward.

The responsiveness of this pH-stat is influenced by both the relative sensitivity of the respective enzymes to changes in pH (steepness of the respective pH curves) and by allosteric effectors of the two enzymes operating in aposing directions. It turns out that dicarboxylic acids activate malic enzyme (when carbohydrate is being metabolized) but inhibit PEP carboxylase, whereas fructose diphosphate activates the latter and inhibits the former (when CO_2 is being fixed (Bonugli and Davies, 1977). The allosteric properties of PEP carboxylase from Avena coleoptiles appear consistent with a role in a cytoplasmic pH buffer when viewed with current theories of IAA-induced growth, wherein enhanced growth is due to the secretion of H^+ into the cell wall solution (Davies, 1980). IAA has been shown to stimulate CO_2 fixation (dark reaction) producing an increase in malate, which is stoichiometrically linked to K^+ uptake (Haschke and Luttge, 1975, 1977) without having a direct effect on PEP carboxylase (Stout and Cleland, 1978).

Hence, any external factor affecting either the energy-dependent extrusion of H^+ coupled to K^+ uptake, or having a direct effect on the enzymology of such pH-stats, could produce an extreme change in cytoplasmic pH and seriously affect many enzymes; activating some and inhibiting others. Daly has argued persuasively that continuous interference in the electrochemical balance of membranes would affect the integrity of pH-stat regulated cytoplasmic pH; one effect of which would be either an increase or decrease in dark CO_2 fixation (Daly, 1980). Toxins acting as ionophores could achieve this result as Daly describes. However, other possibilities exist through regulatory or substrate site-specific effects on key enzymes involved in the generation or decarboxylation of strong organic acids, especially in localized regions of the cell or in specific organelles.

As noted earlier, concentration changes in either positive or negative effectors (including substrates) of allosteric regulatory enzymes, regardless of the primary site or basis of the change, can serve to perturb interlocked systems. It is important to reiterate that the pH-stat just discussed is under allosteric regulatory control by several interactive effectors; all of which could lead to other changes through altered coordination of the malic enzyme:PEP carboxylase interaction. A common consequence of perturbation at this point of metabolic balance would be an observed change in dark CO_2 fixation. Daly (1980) notes that four host-specific toxins including HC-toxin (Kuo and Scheffer, 1970), HS, PC, and HV-toxin (Daly, unpublished), and T-toxin (Bhullar et al., 1975; Daly and Barna, 1980) cause changes in dark CO_2 fixation. Similarly, the oft cited relationship between IAA and fusicoccin, sensu possible common modes of action, is perhaps instructive for comparative biochemistry with certain pathotoxins but should not limit the focus of future studies.

d. Regulation by light. A different but perhaps interrelated control over the activity of many enzymes in plant cells has been illuminated in the past decade or so;

namely the light-dependent regulation of enzyme activity. It is clear that physiological changes induced in plants by several pathotoxins are dependent on or are affected by light. Since the exact relationship between light and the mode of toxin action is, in most cases, unclear it seems appropriate to briefly consider known mechanisms by which light could affect metabolic regulation at the enzyme level.

Evidence now indicates that the carbon reduction phase of photosynthesis, (dark CO_2-fixation) requires light for activation of a number of enzymes that are sparingly active in the dark. Further, the list of enzymes under light control includes not only enzymes involved in the reductive pentose phosphate (RRP) cycle of photosynthesis but also enzymes with diverse roles in secondary plant metabolism, sulfate assimilation, and dicarboxylic acid adduction (Buchanan, 1980). The important point is that light-dependent regulation of these enzymes involves a post-translational alteration in enzyme activity. The regulatory function of light encompasses both enzyme activation (RPP cycle) and enzyme deactivation (starch degradation, glycolysis, and oxidative pentose phosphate pathways). This interaction provides a mechanism to permit competing pathways (anabolic and catabolic) to coexist in the same region and even share common nonregulatory enzymes. Although light apparently acts through the photosynthetic apparatus, the resulting signals pass to and act upon enzymes which are not restricted to the chloroplasts (Buchanan, 1980).

It is currently recognized that several mechanisms exist to mediate light activation of enzyme activity. In the chloroplast, light generated ATP and NADPH affect a structural modification in the key RPP cycle enzyme NADP-glyceraldehyde-3-phosphate dehydrogenase (NADP-GAPD). This enzyme is activated up to fivefold in the light over the rate observed in dark controls at physiological ATP and NADPH concentrations, but only in the presence of P*i* (Wolosuik and Buchanan, 1976). However, P*i* alone has little effect on NADP-GAPD in the absence of activating concentrations of ATP and NADPH.

Similarly, enzymes with rigid concentration requirements for specific ions such as mg^{2+} or an alkaline pH (Racker and Schroeder, 1958) respond to factors which change intercellular concentration of specific ions. The light-regulated flux of H^+ and mg^{2+} if altered by a membrane active pathotoxin (Daly, 1980) could either directly or indirectly affect the activity of several enzymes (Buchanan, 1980). For example, an interesting interaction between mg^{2+} and pH was shown by Priess et al. (1967) where increased concentrations of mg^{2+} decreased the pH optimum of fructose-1,6-bisphosphatase. Examples such as the above, while they may not define the site of toxin action, may explain the mode of toxin action in altering key functions such as CO_2 assimilation.

Another light-regulated mechanism with a possible relationship to pathotoxin induced alterations in enzyme activity involves the protein-mediated reversible activation-inactivation of several enzymes. Current evidence indicates that electrons from chlorophyll are transferred to soluble ferrodoxin in the light, and then to thioredoxin in a reaction catalyzed by thioredoxin reductase (Wolosiuk and Buchanan, 1977). Reduced thioredoxin in turn activates five enzymes in primary carbon metabolism in addition to also activating phenylalanine ammonia lyase. Alternatively, there is evidence for a chloroplast membrane bound reductive system referred to as "light effect mediators" (or LEMs) which may replace the thioredoxin requirement (Anderson et al., 1978).

In the dark, deactivation of the light-activated enzymes occurs, presumably by reversal of the above activation mechanisms although this has not been fully described. Experimental evidence suggests that the deactivation rates are slower than the activation rates. The physiological significance of light-mediated control of activation-deactivation rests with the fact the enzymes of the reduction pentose phosphate cycle (synthesis) appear to intermix with enzymes of the oxidative pentose phosphate and glycolytic pathways (degradative) inside the chloroplast (Buchanan, 1980). In this regulation scheme the two potentially competitive (therefore "futile") cycles are separated.

It is obvious that chemically active pathotoxins capable of interferring with either light-dependent activation or deactivation would soon alter the metabolic steady state by permitting such futile cycles to occur. Tabtoxin, produced by _Pseudomonas_ _syringae_ pv _tabaci_, was reported to inhibit ribulose 1,5-bisphosphate carboxylase (Crosthwaite and Sheen, 1978) which is light activated by an ion-mediated mechanism (Paulsen and Lane, 1966). Tentoxin is reported to interact with chloroplast coupling factor CF_1-ATPase. The latter contained a thioredoxin-like component in the CF_1 fraction enriched in the "S" subunit thought to be essential for attachment of CF_1 to the chloroplast membrane (Binder et al., 1977).

The effect of light on toxin induced necrosis, as well as dark and light CO_2 fixation, has been reported for T-toxin (Bhullar et al, 1975). Preincubation of leaf disks in light increased the sensitivity of dark fixation to T-toxin presumably by some effect on PEP carboxylase, although no direct effect was observed on the enzyme *in vitro*. However, a light mediated effect on stromal pH could lead to changes in the conformational state of PEP carboxylase *in vivo* and provide enhanced sensitivity to T-toxin; a hypothetical scenario which remains a possibility for leaf tissue since the enzyme apparently was not assayed in the presence of both T-toxin and PEP-carboxylase effectors (Bhullar et al, 1975). The physiological status of the PEP carboxylase pH-stat in roots, though presumably not responding to light-dependent pH changes, could have been influenced by other external factors in the enzyme environment, producing conformational states of this regulatory enzyme equivalent to those existing in the light and lead to an analogous sensitivity to T-toxin.

There are many factors that should be considered in evaluating possible pathotoxin-enzyme interactions. Especially vulnerable are allosteric regulatory enzymes which are usually composed of multiple, interactive, protein subunits and which surely function *in vivo* in the presence of their allosteric effectors (both activators and inhibitors). Critical evaluation of pathotoxin binding potential, and subsequent effect on catalytic ability of such enzymes, requires careful assessment of the allosteric state(s) at the time of assay. Serious errors may ensue if attempts to determine effects of pathotoxins on enzymes are conducted without measurements over a known range of regulatory states. Just as the presence and relative concentration of allosteric effectors modifies the binding of each to the enzyme so might they modify the binding of a toxin. Conversely, the binding of a toxin may not affect the binding of all ligands of the enzyme but only some. If the key one is not the substrate itself then the *in vitro* assay would be unrevealing.

Therefore, it is crucial that regulatory enzymes be assayed *in vitro* in both regulated and deregulated states with combinations of their various effectors and the pathotoxins. In the case of light-regulated enzymes this precaution presumably would include retention of the ability of the enzyme to undergo activation-deactivation transitions in the presence of the toxin. Lastly, the possible presence of isozymes of potential toxin targets, localized in the cell but released together in cell free preparations, leads to further caution in interpreting negative data from assays where the various isozymes are not separated and assayed individually.

e. Regulation by compartmentation. The existence of special metabolic functions unique to a given organ is an easily recognized form of compartmentation. Herein, the term compartment is used in reference to an organelle or a site within a cell where a particular metabolic system resides. Location of enzymes in certain cellular organelles is another widely recognized form of metabolic regulation (Oaks and Bidwell, 1970). It is sufficient to point out that the systems for the tricarboxylic acid (TCA) cycle, electron transport and oxidative phosphorylation reside in the mitochondria, the glycolytic enzymes occur in the cytosol, and the photosynthetic apparatus occurs in the chloroplast. Moreover, the chloroplast is the site of many important reactions such as sucrose and starch synthesis (Preiss, 1982), nitrite reduction, and conversion of ammonia to glutamate and glutamine (O'Neal and Joy, 1973; Magalhaes et al., 1974) as well as the photosynthetic fixation of carbon dioxide (Latzko and Gibbs, 1969). Less obvious, but readily detectable biochemically, is compartmentation such as that in seeds with high fat content wherein microbodies with enzymes of the glyoxylate shunt are produced during germination (Beevers, 1969).

Organelles provide centers for localizing certain metabolites and cofactors as well as enzymes. Thus ADP, ATP, pentose phosphates, 3-phosphoglycerate, and triosephosphate move freely across the chloroplast membrane. In contrast, pentose and hexose diphosphates, hexose monophosphates, the pyridine nucleotides, and certain of the Calvin cycle intermedaites occur principally in, and do not move freely from, the chloroplast (Heber and Santarius, 1965; Latzko and Gibbs, 1969). Separation of substrates from catabolic enzymes is another form of regulation that is easily recognized. In many plant tissues nucleases, proteases, and carbohydrases exist which, if not physically separated from their substrates in vivo, would wreak havoc upon metabolic processes and could eventually cause death of the cell. Such loss of compartmentation clearly occurs when plant cells undergo toxin induced necrosis although the events leading to the loss in compartmentation is unclear.

The integrity of compartmentation is maintained at the organelle level by the semipermeable membrane surrounding the organelle. Numerous laboratories working with pathotoxins have focused attention on the effect of toxins on the membranes as sites of primary molecular interaction. The role of the plasma membrane in insuring selectivity of ion and metabolite uptake/extrusion is equally well recognized at the cellular level. Unfortunately, no specific membrane localized site of toxin action has been characterized even though electrolyte leakage through membranes of toxin treated tissue has been observed at an early period after toxin administration.

The recent studies of Dunkle and Wolpert (1981) provide a clear reason for caution when using electrolyte leakage to focus toxin-specific site studies on membranes or when concluding that such leakage is an essential requirement for the resulting disease symptoms. HC-toxin (chapter 1) was shown to cause rapid electrolyte leakage from excised tissues of susceptible sorghum genotypes (Gardner et al, 1972) which then gave rise to the hypothesis that the plasma membrane was a primary site of action for this toxin (Gardner et al., 1974). However, when susceptible sorghum seedlings were pretreated with low toxin concentrations which elicited no symptoms and minimal electrolyte loss, and then retreated 10 hr later with toxin concentrations sufficient to cause severe disease symptoms, the symptoms appeared at the same time as unpretreated controls but were not accompanied by marked electrolyte leakage (Dunkle and Wolpert, 1981). These data indicated that symptoms and electrolyte leakage were not causally related and suggested to these workers that, although toxin recognition may occur at the cell surface with attendant loss of electrolytes under some conditions, other physiological effects lead to the characteristic disease symptoms.

Suffice it to simply agree with Daly who stated, "at this point in the development of research on pathotoxins it seems appropriate to undertake such investigations (membrane studies) when two conditions exist: a purified pathotoxin and a biological basis for believing membrane properties are altered." However, the key role played by organelles through their limiting membranes in contributing to maintenance of the metabolic steady state, as well as their noted sensitivity to alteration by various chemical and physical factors makes them a logical toxin-target.

IV. Experimental Standards and Problems

A. *Experimental Prerequisites*

1. Toxin purity. There are a number of aspects that have proven troublesome in reaching final conclusions in mode-of-action studies on pathotoxins. For example, a number of fundamental physiological processes are affected in plants treated with pathotoxins, some of which may represent important clues to potential molecular sites of interaction. Given the scope and number of such alterations in normal host physiology, or in some cases non-host plants, it seems intuitive that one prerequisite of singular importance is the purity of the toxin preparation used in studies of molecular modes of action of pathotoxins. In most every instance where conclusive evidence for a molecular mode of action has been obtain, the toxin preparation (at least for the key molecular studies) has been of high purity. In contrast an impressive number of descriptive studies over many years have failed to define a molecular site of action for HV toxin, which as yet is uncharacterized. The accumulated efforts with victorin have been reviewed many times (Scheffer, and Yoder, 1980 and Yoder, 1981; Daly 1981 and 1982) with the most recent (Daly, 1981) concluding strongly that further mode of action studies with victorin be postponed until preparations of demonstrated purity are available.

There are a number of reasons why toxin purity is an overriding concern in the design and interpretation of studies directed to the molecular level. The first arises directly from the intent of such studies: a molecular explanation of toxin action. Since pathotoxins used are, in nearly all cases, obtained from cell-free culture filtrates of the pathogen the presence of both media constituents and products of cellular metabolism (either liberated from living cells or released through autolysis) in the filtrates is assured. While phytotoxicity unrelated to the pathological role of components of the culture filtrates is not assured, the concern certainly should be.

As Daly and coworkers (Daly, 1981, and Daly and Knoche, 1982) have correctly noted the multiple and diverse effects often observed for many toxins complicate the notion that there is a single site of action though there well may be only one site. These multiple effects have been noted at the physiological level such as with T-toxin produced by *Helminthosporum maydis* where stomatal closure (Arntzen et al., 1973) dark CO_2 fixation (Behullar et al., 1975), photosynthesis (Bhullar et al., 1975) and ion leakage (Payne et al., 1980) all appear sensitive when individually tested. In addition, structural characterization of several pathotoxins indicates that mixtures of toxin forms occur as with PC-toxin (Wolpert and Dunkle, 1980) AM-toxin (Nishimura et al, 1979) T-toxin (Kono and Daly, 1980) and AAL toxin (Bottini and Gilchrist, 1981). The above observations speak directly to part of the complex problem associated with studies of the molecular mode of toxin action. Namely, that the physiological impact of pathotoxins may involve multiple targets for multiple toxins which are potentially confounded with other components present in impure preparations of pathogen culture filtrates.

2. *Toxin Characterization.* Complete and accurate knowledge of the structure of pathotoxins should be a key prerequisite to the design and interpretation of molecular mode of action studies. Others have commented (lamented) on the fact that so much work on mechanisms of action has preceded structural characterization of biologically active compounds (Daly, 1981). It is intuitive that this prerequisite is closely linked to the need for sufficient quantities of chemically pure toxin preparations; the two really become one overriding goal before extensive biological studies can be considered. These limitations arise partly out of a retrospective view that the most complete understanding of pathotoxin molecular action has come from systems where the toxin structures were known.

Structural information guided the original work on tabtoxin by Wooley et al. (1952) which pointed to methionine metabolism as a possible general site of action. However, the structure was incorrect as originally proposed (Taylor et al., 1972) and this likely compromised the original choice of target sites. Preliminary studies of enzymatic reactions sensitive to Rhizobitoxine were made based on the structural relationship of known inhibitors of pyridoxal phosphate dependent enzyme reactions (Owens et al., 1968; Giovanelli et al., 1971), a choice which appears to have successfully detected inhibition of β-cystathionase (Giovanelli et al., 1973).

There are a number of good reasons why accurate structural information should facilitate subsequent studies designed to relate the structural characteristics of the toxic compound to its molecular role in inducing disease stress. Structural analogies between the pathotoxin and intermediary metabolites in the host plant constitute a basis for focusing attention on specific metabolic sequences or physiological parameters. Various chemical modifications including derivatization, elimination of specific groups, specific addition of radioactive or fluorescent labels, and synthesis of structurally related compounds become possible; any or all of which may provide important clues into primary effector sites. The status of structure-activity relations has been summarized recently by Ballio (1981).

Structural information is necessary also to develop direct chemical and physical assays for pathotoxins. The importance and effective use of chemical or physical assays in concert with biological assays has been reviewed recently (Yoder, 1981). As noted earlier, the main source of pathotoxin preparations is liquid culture filtrates after growth of bacterial or fungal pathogens. Obviously, this is an unnatural media compared to the medium provided by the host plant. Hence, it ultimately becomes important to be able to relate the pathotoxin presence _in vitro_ to the situation as it occurs _in planta_. As noted by Mitchell (1981), it seems both logical and likely that the same pathotoxic compounds will be produced in both situations, but it is not obligatory.

Perhaps even more critical in terms of structural clues for molecular studies is the fact that although the toxins produced _de novo_ in both situations may be the same, the toxin in the plant is exposed to a different set of biological conditions as it moves from its site of synthesis to the site of action. Hence, the final biologically active toxin which perturbs the plants metabolic steady state may be structurally different from the native molecular characterized from culture. Knowledge of toxin structure, structural analogs, and possible metabolites of the native toxin all are potentially useful as clues in this context. Also, direct chemical or radiochemical assays and reliable, efficient purification procedures represent tools for detecting _in vivo_ modification of the native toxin.

Examples of enzymatic modification of toxins by the host plant include phaseolotoxin and tabtoxin produced by _P. syringae_ pv. _pisi_ and _tabaci_, respectively. In the case of phaseolotoxin, (N^S-Phosphosulfamyl) ornithine, the functional toxin _in vivo_ is not produced _in vitro_ but likely represents the enzymatically hydrolyzed product of phaseolotoxin which is produced _in vitro_ (Mitchell and Bieleski, 1977). Tabtoxin appears to require enzymatic modification before interacting with glutamine synthetase, a proposed target site in the host. The active inhibitor has been characterized as tabtoxinine-β-lactam which arises secondarily in the plant by peptidase cleavage of tabtoxin (Durbin 1982).

In conclusion, precise chemical characterization of pathotoxins is a necessary prerequisite to molecular mode of action studies but history reveals that this goal is neither simple to achieve nor are initial structural proposals infallible. A number of proposed structues have required subsequent revision as noted by several reviewers (Mitchell,

1981; Daly, 1981). However, the same authors provide a firm basis for the need to extend knowledge in the area of the chemical nature of pathotoxins; insights to which the reader is strongly directed.

3. Assays and experimental material. Detection of biologically relevant effects of purported pathotoxins rests squarely on the sensitivity, specificity, variability, and reliability of the assay or assays available. The ability of assay procedures to probe the determinative molecular events depends still further on the choice and handling of physiologically competent host tissue. Bioassays initially define the phytoxicity of a pathotoxin while chemical assays, based on toxin structure, complement and hopefully facilitate other aspects such as purification and in vivo detection of pathotoxins or their metabolites. Yoder (1981) has reviewed various approaches taken by pathotoxin coworkers in developing assay procedures. This review is comprehensive in terms of the criteria used to develop assays to detect biological activity and to assess the role(s) of pathotoxins in disease. Thus, only a few points need to be stressed in terms of either assays or choice of experimental material in relation to studies at the molecular level of toxin interaction.

The molecular site of action will be determined, in the final analysis, from results of bioassays. This may (likely will) require not one but several different bioassays. For instance, the mode of action of phaseolotoxin was defined using bioassays for chlorosis, (Woolley et al., 1952), the accumulation of ornithine and its reversal by arginine and citrulline (Patil et al., 1972), and finally assay of ornithine transcarbamylase (Mitchell, 1979). Where possible, it is important also to use more than one type of assay since different assays are known to give conflicting results (Yoder, 1981). In terms of assays involving specific target sites such as in enzyme or membrane preparations, it is important that the assays be conducted under several limiting physiological conditions (temperature, pH, ionic strength); that the interaction of metabolic effectors be considered; that metabolically important parameters (such as regulatory properties) be monitored throughout the experimental period; and that each of these be related back to the physiological impact of the toxin in intact tissue.

The choice of host tissues also should receive strong consideration. Age-structure differences in relation to relative pathotoxin sensitivity are well documented (Yoder, 1981). Therefore, the tissue source should reflect biological sensitivity to the toxin and should, when possible, be consistent with the type of tissue normally associated with symptom expression. Moreover, morphogenetic differences in cell types may be important if one wishes to assess the effect of a toxin on a particular physiological process which is localized in certain cells but not others. For example, assessment of a suspected pathotoxin effect on PEP carboxylase should include awareness that the enzyme extracted from roots of several C-4 plants is distinct from the leaf forms (Ting and Osmond, 1973). This same enzyme can be used to illustrate another hypothetical pitfall in terms of assuming that all preparations of an enzyme from the same tissue reflect a single functional state. PEP carboxylase from crassulacean acid metabolism (CAM) plants, showed a 12-fold lower Km for PEP during a dark period than a light period (O'Leary, 1982). However, if extracts obtained during a dark period were allowed to stand for an hour or more prior to assay the "dark" form was converted to the "light" form due to a proposed covalent modification (Winter, 1980). Both forms show significant catalytic activity but they differ in a number of kinetic properties.

Lastly, the genetic variation in host plants used as tissue sources is a factor to consider strongly. In the final analysis the use of isogenic lines or clones should reduce the genetic component of variability to a minimum.

V. Proposed Molecular Sites of Action

Given the number of pathotoxins described since 1947 it is surprising and no doubt frustrating to many workers that data supporting unique molecular sites of interaction have been forthcoming for so few. The lack of structural information may explain a significant part of this notable lack of success. As emphasized in the preceding discussion, definition of a molecular mode of action is dependent on knowledge of the molecular structure of the pathotoxin probe. A number of seemingly very good clues from physiological changes induced by a number of pathotoxins with different host systems have been inconclusive as impirical predictors of primary target sites. In retrospect, it may be that there were too many "good" clues without a reliable basis to eliminate those due to secondary consequences of one or more primary sites of interaction. Others have extensively reviewed the literature dealing with studies on plant tissue or cell responses to toxins for clues to the molecular mode of action (Daly, 1981; Durbin, 1982; Daly and Knoche, 1982). These authors have discussed possible sites where either the data collected or the experimental methods used were inadequate to directly implicate a single site or even confirm that the physiological response measured was uniquely indicative of the primary target site.

The remainder of this discussion will review examples of chemically specific interactions of pathotoxins with host-gene. In all cases reported so far, experimental results reveal alterations in specific enzyme mediated functions which first were deduced through in vitro studies. Presumably, the resulting change in catalytic activity at these proposed sites, if it occurs in vivo, can be amplified through an integrated metabolic network, described earlier, in order to induce the detrimental physiological impact characteristic of each pathotoxin symptom.

A. *Rhizobitoxine*

Certain strains of Rhizobium japonicum (=melitoti) produce a pathotoxin in root nodules colonized by these normally symbiotic bacteria. The first recovery of this presumptive toxin was directly from nodules on soybean plants which showed unusual chlorosis in young developing leaves (Johnson et al, 1959). Water extracts of nodules from chlorotic plants induced chlorosis not only in seedlings of the soybean cultivar from which they were recovered but other non-host plants as well. Owens and Wright (1965a) subsequently isolated sufficient quantities of the toxin from nodules to begin structural characterization. They determined the toxin concentration in the nodules, and isolated it from chlorotic young leaves but not older tissue. These workers next developed a synthetic medium to culture pure strains of R. japonicum from which they were able to obtain relatively large amounts of the phytotoxin for further structural characterization (Owens and Wright, 1965b). Taken together these results provided the beginning of a desirable scenario to focus on the molecular mechanism(s) of toxin action. The toxin was produced in vivo, recovered from tissue showing disease symptoms, and structural characterization was begun with large quantities of the toxic compound which was later given the trivial name, rhizobitoxine.

The isolated active compound was shown to inhibit growth of several other organisms including Chorella and Salmonella typhimurium (Owens et al, 1968). Various compounds were added to the growth medium of S. tryphimurium to attempt to

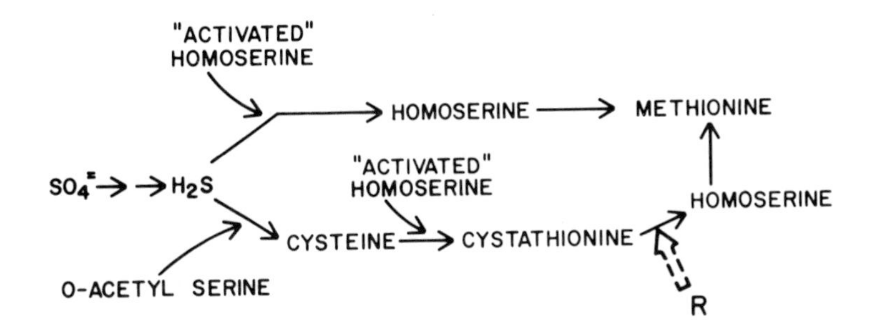

FIGURE 3. Pathway of methionine biosynthesis. The proposed site of action for rhizobitoxine is indicated at the step catalyzed by β-cystathionase (dashed arrow, R).

antagonize (relieve) the growth inhibition due to rhizobitoxine. Inhibition was overcome by methionine and homocysteine but not by serine, homoserine or cystathionine (Owens et al, 1968). The methionine biosynthetic pathway, previously characterized in bacteria and fungi (Flavin, 1975), provided a possible explanation for these results (Fig. 3).

In sequence, o-acetyserine, cysteine, homoserine and cystathionine are precursors of homoserine which is then converted to methionine. This sequence includes transulfuration reactions in which the sulfur atom of cysteine is transferred to homoserine via the thioether cystathionine. Thus, the metabolic lesion induced by rhizobitoxine appeared to exist between cystathionine and homoserine, a reaction catalyzed by cystathionine-β-lyase (EC 4.4.1.8), commonly known as β-cystathionase.

This hypothesis was supported further when cell free preparations of the bacterial enzyme were inhibited *in vitro* by purified rhizobitonine. Although these results remained to be confirmed with a plant system both *in vitro* and *in vivo* they illustrate two more steps in the evolving scenario. First, Owens and coworkers developed a non-host assay system of toxin activity which utilized a rapidly growing organism (*S. typhimurium*) which was sensitive to the toxin and could be grown in a chemically defined medium. Secondly, the use of biochemical ammendments in the defined medium provided an opportunity to probe specific metabolic pathways for possible clues to molecular sites of action if specific intermediates of a pathway could antagonize the action of the pathotoxin. In theory, intermediates proceeding the site of inhibition would not relieve the toxin effect but those following the altered step would restore normal flux through the remainder of the pathway in the presence of a site-specific antimetabolite.

The structure of rhizobitoxine was reported in 1972 as 2-amino-4-(2-amino-3-hydroxypropoxy)-trans-but-3-enoic acid which belongs to a rare group of natural products including cystathionine, the enol-ether amino acids (Owens et al, 1972).

Rhizobitoxine: $HOCH_2CH(NH_2)CH_2OCH = CHCH(NH_2)COOH$

Cystathionine: $HOOCCH(NH_2)CH_2SCH_2CH_2CH(NH_2)COOH$

The enol-ether amino acids contain several representatives which are antibiotics (Mitchell, 1981). However, it remained to be demonstrated that the _in vitro_ bacterial results were functionally linked to a similar site of action in plants. As a first step, β-cystathionase, isolated from spinach leaves, was shown to be irreversibly inhibited by the purified toxin (Giovanelli et al, 1971). The results revealed also that spinach β-cystathionase was inactivated by the toxin by an active-site-derived irreversible inhibition which, in enzymological terms, has the mechanism:

$$E + R \rightleftharpoons EPR \longrightarrow EPR_I$$

[where EP = the holoenzyme form of β-cystathionase (P, pyridoxal phosphate); R = rhizobitoxine; EPR = a dissociable enzyme - rhizobitoxine complex; and EPR_I = an inactivated enzyme]. The inhibition was not reversed by removal of free (unbound) toxin but was prevented by the substrate cystathionine. Interestingly, pyridoxal phosphate (a cofactor) specifically reactivated the enzyme. Further studies addressed three key questions:

1) Is β-cystathionase inhibited _in vivo_ as well as _in vitro_?
2) Do alternate pathways exist to compensate for the metabolic block?
3) Does inhibition of β-cystathionase _in vitro_ account for the pathological effects of rhizobitronine?

To answer the first question rhizobitoxin-treated corn seedlings were allowed to assimilate $^{35}SO_4$ for 3 or 6 hours and the radioactivity incorporated into sulfur amino acids was assesed (Giovanelli et al, 1972). The most striking effect of the toxin was a 22 fold increase in radioactive cystathionine, presumably due to blockage of the β-cystathionase step. However, accumulation of radioactive methionine was only slightly inhibited, implying an alternate route of methionine synthesis was available. Biochemical evidence for a second route from SO_4 to methionine by direct sulfhydration of homoserine to form homocysteine does exist in plants. While these results do not reveal the relative contribution of the two possible pathways in the absence of the toxin they do indicate that the pathway involving β-cystathionase does contribute to methionine biosynthesis and that metabolism via this pathway is perturbed by rhizobitoxine.

The data do not appear to explain why inhibition of β-cystathionase does not exert a larger effect on methionine synthesis. The answer may involve a secondary consequence resulting from cystathionine accumulation due to inhibition of β-cystathionase. Giovanelli et al. (1972) considered the fact that cystathionine concentrations in normal corn seedlings were at least 100-fold less than the in vitro Km for β-cystathionase. This indicated that in vivo the enzyme operated well below substrate saturation. Thus, if inhibition of β-cystathionase by the toxin raised the concentration of the substrate (observed 22-fold) the residual enzyme activity could operate at a relatively higher flux (nearer the Km) and serve to restore normal pathway function. Should this be the case, it would be important to study the kinetics of ^{35}S labeling at earlier time periods.

These findings alone to not support a sole-site hypothesis via β-cystathionase inhibition since methionine synthesis was not totally inhibited. In keeping with the caution discussed earlier regarding age structure differences in enzyme activity and/or regulatory patterns, the methods used by Owens and coworkers did not assess tissues by development age. They, in fact, state that only the basal areas of young leaves used in their studies showed chlorosis while their extracts also included tissues which did not show overt symptoms (Giovanelli et al., 1973). Hence, it is possible that localized methionine difficiencies could have been obscured by the choice of tissue.

While this step is a potential site of toxin interaction there are other sites in the cell where methionine plays a significant role. One such site, subsequently shown to be affected by rhizobitoxine and several analogs, is at the enzymatic conversion of methionine to ethylene (Lieberman et al, 1979). Adams and Yang (1979; Boller et al, 1979) have demonstrated that an analog of rhizobitoxine inhibits the conversion of S-adenosylmethionine to 1-aminocyclopropane-1-carboxylic acid, a precursor of ethylene.

In summary, the results to date indicate at least one specific enzyme site in the cell where rhizobitoxine binding alters catalytic activity in a primary metabolic pathway. However, the existence of alternate, biologically significant, sites remains to be full characterized both in vitro and in vivo. These studies do provide a model experimental scenario which may be applicable to other systems.

B. *Tentoxin*

As a general symptom, chlorosis ranks high on the list of consequences of toxin activity. Like rhizobitoxin, the phytotoxic cyclopeptide tentoxin, produced by Alternaria alternata (= tenuis), causes chlorosis on a number of plants, including cotton and citrus seedlings which are hosts for the pathogen. The disease was first characterized by Fulton et al. (1960) who reported that the chlorotic condition of cotton seedlings from which A. alternata was routinely isolated could be reproduced with sterilized culture filtrates of the fungus. The disease was recognized by a distinctive varigated, irreversible seedling chlorosis on cotton where yellow chlorotic areas were sharply delineated from apparently unaffected green areas on cotyledons. Once affected, seedlings remained chlorotic, grew at reduced rates, and usually died if the chlorotic areas covered more than 35% of the total cotyledondary area (Fulton et al, 1965).

Templeton and coworkers (Fulton, 1965) evaluated a number of factors which affected the choice of bioassay tissue and conditions affecting the sensitivity of assays to the purported toxin. Among these were the sensitivity of various plants (i.e. the toxin host range) the effect of tissue age-structure, the role of light, and the toxin exposure time required to induce a concentration dependent chlorotic response. The thoroughness with which these early studies considered the limiting biological and environmental factors ultimately provided not only a suitable assay system for the toxin, but also important clues about the mode of action.

The pattern of stress development in which some, but not all, cells in cotton cotyledons exhibiting chlorosis presented a problem due to the lack of uniformity in cellular responses. Cucumber seedlings were selected to avoid this problem since they were quite sensitive to the toxic factor and chlorosis was more uniform than in cotton (Templeton, 1972).

Seedling-age structure was revealed to be a second significant factor determining relative tissue sensitivity to the toxin. Plants were sensitive only during the first 32 hours of growth of the seedlings prior to or during the initiation of greening in the developing seedling. The response of different ages of tissue to tentoxin was determined by exposing the seeds or seedlings to a 1 hr toxin treatment (0.3 mg/ml) after 1, 16, 24, 32, and 48 hours germination in the light. Only those seedlings exposed to the toxin at 32 hours or earlier developed chlorosis. Seedlings treated at 48 hours after germination began remained green and did not show visible evidence of toxin induced stress. In fact, the hypocotyls of seedlings treated at 32 hours were even light green with a trace of green at the base of the cotyledons. Taken together these observations indicated that for the toxin to cause chlorosis it had to be present in the tissue prior to completion of chlorophlast development. Furthermore, the effect of tentoxin on chlorophyll content of the cucumber seedlings was concentration dependent, as expected from the proceding observations.

Consistent with the emerging concept of a role for the toxin in interfering with normal chlorophyll synthesis was the effect of light on symptom development. The sensitivity of plants to the toxin was increased by holding them in the dark after toxin administration. The amount of chlorosis increased with the increasing length of dark periods across the range of 0, 24, 48, and 64 hours. Complete chlorosis resulted when the seedlings were held in the dark for 64 hours after the toxin treatment compared to untreated control seedlings held in the dark for the same period which developed normal green cotyledons when both were returned to the light. These studies by Templeton and coworkers thus provided important information establishing the critical roles of light, morphogenetic development, and the choice of plant species required to effectively study toxin purification, characterization, and the molecular mode of action.

The toxin was purified and detailed structural information reported almost simultaneously from two laboratories (Koncewicz et al, 1973; Meyer et al, 1974a). Differences between the two reports in the sequence of the amino acids of the cyclic tetrapeptide were resolved and the structure for tentoxin concluded to be cyclo(L-leucyl-N-methyl-(Z)-dehydrophenylalanyl-glycyl-N-methyl-L-alanyl) (Meyer et al, 1974b) structure. Synthetic procedures were developed to unambiguously confirm the structure (Rich et al, 1975) and to prepare various analogs (Rich et al, 1978). These efforts have provided excellent materials for studies of structure-activity relationships to the point where such studies with tentoxin have reached a truly advanced state (Ballio, 1981).

With the availability of a pure and chemically characterized compound, mode of action studies were conducted in several laboratories with the focus on chloroplast development and chlorophyll syntheses. Halloin et al. (1970) examined both chlorophyll synthesis and plastid development and concluded that the site of action lay with proplastid and lamellar formation. The first major breakthrough was reported by Arntzen (1972) who reasoned that since chloroplast development is strongly linked to a high energy requirement for normal grana formation, the direct effect of tentoxin on energy coupled reactions was a likely site of action. The evidence confirmed that cyclic photophosphorylation, but not reversible proton accumulation, was inhibited by tentoxin in isolated lettuce chloroplasts. Further, tentoxin inhibited coupled electron flow in the presence of ADP and phosphate but did not alter the basal electron flow in the absence of these acceptors or in an uncoupled electron transport system. These data pointed to an energy transfer inhibition by the toxin at the terminal steps of photophosphorylation dependent ATP synthesis (Arntzen, 1972). Support for such a hypothesis would include demonstration of the requirement for light driven ATP synthesis rather than the failure of some other step in the synthetic processes leading to normal lamellar development.

Within three years direct evidence for the role of tentoxin-induced inhibition of photophosphorylation in altering normal chloroplast development was provided by Bennet (1975). Both RNA and protein synthesis (ribulosebisphospate carboxylase) in isolated pea shoot chloroplasts were inhibited by tentoxin when the synthetic reactions were driven by light-dependent ATP but not when driven by ATP exogenously supplied in the dark. Along with tentoxin, several inhibitors of RNA and protein synthesis including actinomycin D, lincomycin, D-threo-chloramphenical, and carbomyl cyanide m-chlorophenyl hydrazone (CCCP) were compared using this system. Only CCCP, an inhibitor of phosphorylation, was exclusively active in the light-dependent part of the system and therefore mimicked inhibited by tentoxin (Bennet, 1975). Consistent with these data were the results of ultrastructural studies by Halloin et al. (1970) that the effects of tentoxin were restricted to the chloroplast.

Extension of the above studies to a specific site in the energy transfer system of the chloroplast came from the deduction that chloroplast coupling factor 1 (CF_1) is directly involved in photophosphorylation and thus represents a possible binding site for tentoxin (Steele et al, 1976). As will be discussed shortly, the efforts of the Wisconsin group have demonstrated in an unambiguous fashion that tentoxin binds to the CF_1-ATPase complex in lettuce chloroplasts which results in an inhibition of photophosphorylation and CF_1-ATPase activity. However, before considering these studies a brief description of the nature and role of the CF_1-ATPase complex in the unperturbed chloroplast may be useful.

Light generated electron transfer in chloroplasts creates a proton gradient across the thylakoid membrane where the electron transfer chain is located (McCarty, 1979). Ultimately, the transfer of electrons down the proton gradient provides the energy needed for ATP synthesis. The name for the ATP synthesis enzyme is taken from the reverse reaction, namely ATP hydrolysis (ATPase). Under certain experimental conditions (e.g. treatment with EDTA) a protein(s) can be released from the membrane and ATP synthesis is "uncoupled" from the proton gradient, i.e. no further ATP synthesis occurs. Under other conditions these proteins may rebind to the membrane restoring electron flow and ATP synthesis. Hence, the protein involved in the "recoupling" of ATP synthesis to electron transport is termed a coupling factor (McCarty, 1979).

The coupling factor released by EDTA treatment of chloroplasts was found by McCarty and Racker (1966) to be identical with CF_1 isolated from mitochondria and to contain both a latent Ca^{++}-dependent ATPase and a coupling factor. CF_1, which comprises about 10% of the thylakoid membrane protein, is a colorless water soluble protein which contains tightly, but not covalently, bound ADP and ATP (1:1) along with about 3 moles of carbohydrate per 100 moles of amino acid. The molecular weight is approximately 325,000 and the protein is made up of five electrophoretically distinct polypeptide components which are labeled alpha to epsilon in order of decreasing size as found in a tryptic digest of the native protein (McCarty, 1979). Hence, any factor or factors which displace the CF_1 complex from the membrane or intract directly with it to alter its kinetic response to substrates or effectors would serously perturb the energy-dependent metabolic steady state of the chloroplast and, presumably, could do so in a short period of time.

Incubation of tentoxin with isolated chloroplasts and purified CF_1 from lettuce, a toxin sensitive species, indicated that CF_1 had a single binding site for the toxin. When tentoxin occupies this site both CF_1-ATPase and phosphorylation electron transport are inhibited (Steel et al, 1976). An affinity constant of 1.3×10^7 M^{-1} was estimated from the coupled electron transport rates the same affinity constant was estimated for solubilized CF_1 from the inhibition of trypsin activated ATPase activity.

Interestingly, no inhibition of heat- or trypsin-activated ATPase from radish, an insensitive species, was detected even at concentrations 200-fold greater than that required for 50% inhibition of the lettuce enzyme. Unactivated CF_1 from lettuce directly bound ^3H-tentoxin with an affinity constant of 2×10^8 M^{-1} (measured using continuous ultrafiltration) and it was estimated to have 0.81 tentoxin-binding sites per molecule of CF_1. Toxin binding to heat activated CF_1 from lettuce was estimated to be 2×10^8 M^{-1} with 0.85 sites per CF_1 molecule. Neither ATP nor adenyl-5'-yl imidodiphosphate were observed to interact with this site indicating that toxin and nucleotide sites were different (Steel et al, 1976). Similar studies with radish CF_1 gave an affinity constant estimate of 1×10^4 M^{-1}. Thus there appeared to some species-level selectivity in the inhibition of CF_1 by tentoxin.

Later studies confirmed that the tentoxin inhibitor of lettuce CF_1-ATPase was uncompetitive (Steele et al, 1978) and involved binding to a single site on the α- and/or β-subunit with an apparent Ka of 2×10^8 M^{-1} following trypsin digestion of CF_1-ATPase to release a Ca^{+2}-dependent ATPase (Steele et al, 1977). Since neither nucleotide substrates, phosphate, or calcium competed with the toxin binding site or affected the apparent affinity constant; the steady-state kinetics best fit an uncompetitive pattern (Durbin and

Steele, 1978). This suggested that the inhibitor binds after an irreversible step following nucleotide binding and that the binding occurs at a site different than the catalytic site.

In a discussion postulating the evolutionary origin of the tentoxin binding site, the species selectivity was addressed (Durbin and Steele, 1978). A key point to note here is that in all cases (sensitive and insensitive species) there appears to have been an evolutionarily conserved catalytic activity which functions similarly in all species in the absence of the toxin. However, the catalytic site is associated with a species specific and equally conserved site which is capable of binding tentoxin. The origin and any alternate role of the later is unknown at this time.

C. *Tabtoxin*

The presence of chlorosis-inducing compounds in culture filtrates of *P. s.* pv. *tabaci*, was recognized as early as 1925 (see Stewart, 1971). However, it wasn't until the 1950's that systematic studies were begun on the "wildfire" toxin with the efforts of Woolley, Braun, and coworkers who proposed both a mode of action and a structure for the toxin (Woolley et al, 1952; Wooley et al, 1955). Unfortunately, both original proposals proved later to be incorrect, but the presence of potent biological activity and useful bioassays, including plant tissue and *Chlorella*, no doubt captured the interest of those who followed with more sophisticated analytical techniques and a broader available background in plant biochemistry. Today both the structures of the toxin forms produced by pv. *tabaci* and substantial data on an enzyme specific target site consistent with the physiological symptoms have been confirmed by several laboratories. A brief summary of the historical sequence of studies with this thoroughly examined system is illustrative of both an acceptable methological approach and some potential inaccuracies which may be reached if certain initial assumptions are incorrect.

Early efforts to characterize the toxin were hampered by instability of the biological activity and the lack of efficient separatory techniques for handling unstable compounds (Stewart, 1971). However, by the early 1970's both difficulties were overcome and two laboratories almost simultaneously reported corroborative structural evidence (Stewart, 1971; Taylor et al, 1972) from mass spectrometry, oxidative degradation, and nmr data.

The active compound was identified by spectroscopic and chromatographic methods to consist of a mixture of two different dipeptides both containing the unusual amino acid tabtoxinine-β-lactam linked either to L-threonine or L-serine. The threonine analog was given the trivial name tabtoxine and the serine analog was named (2-serin)-tabtoxin. Both analogs were produced by pv. tabaci from tobacco, soybean, bean, and tomato; by pv. coronafaciens from timothy, corn, and oats; and by pv. garcae from coffee and all the host plants are sensitive to the toxin (Sinden and Durbin, 1970). In addition to the active toxins an inactive isomer also was characterized and named isotabtoxin (Stewart, 1971).

An acid catalyzed intramolecular translactamizatin reaction occurs in aqueous solution to give a half-life of tabtoxin of about 1 day (Stewart, 1971; Taylor et al, 1972) which accounts for the rapid loss in activity which hampered earlier efforts to purify and characterize the toxins. More recently the β-lactam of tabtoxinine, which itself is biologically active, has been recovered from culture filtrates of pv. tabaci (Durbin et al, 1978). However, the latter compound does not appear to be a primary bacterial product but arises as a secondary product in the medium by peptidase cleavage of native tabtoxin (Uchytil and Durbin, 1980). The possible origin of tabtoxinine-β-lactam in planta by action of non-specific plant peptidases has lead to the apparent resolution of an interesting dilemma arising from mode of action studies and toxin purification.

Like the structural studies, the mode of action efforts also required periodic reinterpretation of existing data. Before the correct structures of tabtoxin were known it was first postulated that these toxins acted as antimetaboltes of methionine synthesis because of the similarity in action of the toxins and methionine sulfoximine (MSO) (Braun, 1955). When tabtoxin was tested against Chlorella, in a chemically defined medium the algal growth was markedly inhibited (Braun, 1950). Addition of yeast extract to the medium prevented the growth inhibition and when components of the yeast extract were tested singularly only L-methionine was effective in antagonizing toxin action (Braun, 1955). Since methionine did not directly inactivate the toxin it was concluded that the toxin interfered with normal methionine metabolism. In part, this idea was strengthened at the time by the belief that MSO affected methionine metabolism and the observation that MSO produced chlorotic lesions in tobacco which resembled those of tabtoxin. Similarities in the effect of MSO and tabtoxin were correct; unfortunately the proposed mechanism of action was not.

Later studies showed that methionine competed with MSO for uptake sites in Chlorella and thus relieved the growth inhibition but did not alter the induction of chlorosis in higher plants. Studies in animal systems suggested that MSO inhibited cerebral glutamine synthetase (Sillinger and Weiler, 1963). In 1968 Sinden and Durbin reported that partially purified preparations of tabtoxin and MSO both inhibited glutamine synthetase from tobacco *in vitro* at concentrations which caused chlorosis *in vivo*. Lineweaver-Burk plots (1/v vs 1/s) indicated that the tabtoxin inhibition was non-competitive which supported the observation that glutamine did not reverse the inhibition at high tabtoxin concentrations (Sinden and Durbin, 1968). Inhibition of rat cerebral glutamine synthetase by tabtoxin was reported to be competitive *in vitro* but irreversibly bound to the enzyme in the presence of ATP and Mg^{2+} (Lamar et al, 1969). However, in further studies with purified tabtoxin and glutamine synthetase preparations *in vitro* inhibition could not be demonstrated (Durbin, 1981). Thus the dilemma: why did partially purified toxin preparations and MSO inhibit the enzyme but the purified toxin did not affect glutamine synthetase?

A decade later the presence of free tabtoxinine-β-lactam in the culture filtrates of pv. *tabaci* was reported (Durbin et al, 1978). Furthermore, like tabtoxin, it also caused chlorosis and an accumulation of ammonia in the treated tissue, both of which were light dependent. These observations supported the notion that tabtoxinine-β-lactam could be the functional toxin *in vitro*. The search for peptidases capable of releasing tabtoxinine-β-lactam from native tabtoxins was successful (Uchytil and Durbin, 1980) in six plant species and eight isolates of tabtoxin producing bacteria. Preliminary evidence indicated that purified tabtoxinin-β-lactam, unlike purified but like unpurified-tabtoxin, inhibited glutamine synthetase (Uchytil and Durbin, 1980).

The scenario for chlorosis now is as follows: bacterial production of tabtoxin is followed by cleavage of the toxins by plant peptidases to release of tabtoxinine-β-lactam. The latter product then inhibits glutamine synthetase in the plant lead to a buildup of ammonia to phytotoxic levels (Durbin, 1981). Therefore, literally two toxins are involved in symptom expression: tabtoxinine-β-lactam from the pathogen and ammonia from the pathotoxin sensitive plant.

Additional support for the causal role of glutament synthetase inhibition in development of the chlorotic symptoms associated with tabtoxin came from the demonstration that tabtoxin inhibited glutamine synthetase *in vivo* within 4 hr;

a time frame which paralleled ammonia accumulation (Turner, 1981). Tabtoxin (0.5 mg ml) infiltrated into tobacco leaves resulted in a detectable (16.5%) reduction of glutamine synthetase activity within 1 hr, which was further reduced 95% in 4 hr, after which there was no further change in enzyme activity through the 10 day sampling period. Consistent with the purported role of glutamine synthetase in ammonia asimilation in plant cells (Miflin and Lea, 1976) is the fact that the apparent inhibition of this enzyme was accompanied by an accumulation of ammonia which started at 3 to 4 hr after tabtoxin treatment when glutamine synthetase had declined to approximately 25% of the untreated control. The relationship between inhibition of glutamine synthetase, ammonia accumulation, and chlorosis become clear after consideration of the physiolgical linkages involved.

Glutamine synthetase occurs throughout the plant and animal kingdom and has been characterized from bacteria, algae, fungi, higher plants, and animals. The biosynthetic reaction:

$$\text{ATP} + \text{L-glutamate} + \text{NH}_3 \longrightarrow \text{L-glutamine} + \text{ADP} + \text{P}i$$

also can be measured in the reverse direction by the so-called transferase assay. The transferase assay, which is ADP dependent, generally gives rates several times higher than the synthetase assay and has been used for most of the tabtoxin studies. The enzyme occurs in all tissue throughout the plant although 60% and 90% of the total activity may be found in the chloroplast (Wallsgrove et al, 1979). Two forms of the enzyme occur in barley with both forms having similar molecular weights but differing in stability, pH optima, and reaction to thiol reagents, and subcellular localization. The form restricted to green tissues exhibits negative cooperative binding with glutamate while the form found in the cytosol shows standard Michaelis-Menten kinetics (Guiz et al, 1979).

In terms of *in vivo* regulation glutamine synthetase appears to be be feedback regulated by nucleotides including ADP, 5'-AMP which indicates the possibility of control by energy charge. Since the reaction catalyzed by glutamine synthetase is regulated as the first step in a complex of pathways leading to synthesis of amino acids, it is not surprising that several amino acids show regulatory inhibition. When compiled from several plant systems, the amino acids which are inhibitory include alanine, glycine, serine, and aspartate. However, based on apparent inhibition constants and results with isotope kinetic methods (Sims and Fergson, 1974), it appears that regulation of this enzyme is primarily through availability of enzyme substrates, particularly ammonia and glutamate.

Interestingly, carbamyl phosphate is inhibitory to the pea leaf enzyme *in vitro* (O'Neal and Joy, 1975). Whether or not these *in vitro* results are consistent with *in vivo* function is not clear. However, it is noted for future reference that a buildup of carbamyol phosphate due to inhibition of an enzyme normally using it as a substrate could conceivably affect the normal flux through glutamine synthetase.

A second physiologically significant aspect of this enzyme is its apparent role in reassimilation of ammonia produced during photorespiration. Keys et al. (1978) proposed that glycolate produced from protein catabolism during photorespiration is converted to glycine and then to serine + NH_3 + CO_2. The ammonia thus released originally came from glutamate and is reassimilated by glutamine synthetase in the leaf to complete a closed cycle for ammonia detoxification. The flow of carbon through this cycle requires ATP and NADPH produced during photosynthesis. Inhibition of glutamine synthetase by MSO in an illuminated culture of *Chlomydomonas* resulted in ammonia accumulation produced during photorespiration (Culimore and Sims, 1980). Therefore, glutamine synthesis is a key enzyme in the plant nitrogen cycle which is driven by the photorespiratory carbon cycle and coupled to protein catabolism. Inhibition of glutamine synthetase under light conditions presumably would result in an accumulation of ammonia. Normally, the high Km of an uninhibited glutamine synthetase for ammonia would prevent sufficient accumulation of ammonia to uncouple photophosphorylation (Miflin and Lea, 1976).

Further similarities between MSO and tabtoxin have been used to strengthen the argument for a common mode of action. Ammonia accumulation and a corresponding reduction in the rate of photosynthesis were observed in spinach leaf tissue treated with MSO (Platt and Anthon, 1981). This effect, observed in the absence of significant nitrate reduction or exogenously supplied ammonia, was dependent on the O_2 and MSO concentrations supplied to the tissue. Two recent reports link a purported tabtoxin inhibition of glutamine synthetase *in vivo* to light-dependent ammonia accumulations (Frantz et *al*, 1982; Turner and Debbage, 1982) and suggest that accumulated ammonia inhibits photophosphorylation. In both reports the bulk of ammonia generated could be accounted for by photorespiration.

Conditions leading to reduced photorespiration (low O_2 or absence of light) reduced the amount of ammonia accumulated in tabtoxin treated tissues. Also, in both cases, measured decreases of 94-98% were observed for glutamine synthetase in the toxin treated tissues. Thus, the case is quite strong for direct tabtoxin inhibition of glutamine synthetase, probably through peptidase liberation of tabotoxinine-β-lactam *in planta*, leading to an accumulation of ammonia (derived from photorespiration) to a level theoretically high enough to uncouple photophosphorylation and eventually, by steps yet unclear, to lead to chlorosis (Frantz et al, 1982; Turner and Debbage, 1982).

D. *Phaseolotoxin*

Suggestive evidence for alteration in a specific biosynthetic pathway associated with chlorosis in bean plants infected with *P. s.* pv. *phaseolicola* was reported before the structure of the putative toxin was known. Infection of susceptible beans with pv. *phaseolicola* or injection with culture filtrates of the pathogen consistently resulted in an unusual increase in ornithine content in the chlorotic tissue and probably other compounds as well (Patel and Walker, 1963; Rudolph and Stahman, 1966). Increases up to 100-fold in ornithine content of toxin-treated tissues were particularly significant because normal pool sizes of this compound are small due to tight regulatory control via the ornithine cycle, or by conversion to glutamic semialdehyde and 2-oxo-5-amino valeric acid (Thompson, 1980). Furthermore, the ornithine increases did not occur in cells which were rendered chlorotic by other means.

Chlorosis-inducing, partially purified toxin preparations from a virulent strain of pv. *phaseolicola* inhibited red kidney bean ornithine carbamoyl transferase (OCTase, E.C. 2.1.3.3) *in vitro* (Patil et al, 1970). Similar toxin preparations had no effect on glutamine synthetase, glutamine transferase, carbamyl phosphate synthetase, aspartate carbamoyltransferase, or arginase at toxin concentrations 100-fold greater than those that inhibited OCTase by 50% (Patil et al, 1972). Significantly, L-citrulline and L-arginine-HCL protected against and reversed toxin-induced chlorosis. Accumulation of ornithine, obsrved earlier by other workers, did not occur under these experimental conditions but was achieved in later studies by these workers and others.

Preliminary studies with 10-fold purified OCTase preparations recovered from an acetone powder of red kidney bean revealed inhibition kinetics consistent with competitive inhibition by the toxin for carbamyl phosphate but noncompetitive for ornithine. In contrast, OCTase activity measured in cell-free extracts of susceptible bean cultivars inoculated with virulent isolates of pv. *phaseolicola* was significantly inactivated compared to similar preparations obtained from resistant plants in which no toxin was produced (Gnanamanickam and Patil, 1976). Since the inactivated preparations were dialyzed before *in vitro* assay, it appeared that the toxin binding *in vivo* led to irreversible inactivation; at least the toxin could not be easily solublized away from the enzyme.

Additional studies designed to apply the Meloche (Meloche, 1967) model for a noncovalent enzyme inhibitor complex indicated that OCTase inhibition by partially purified toxin preparations could not be reversed by dialysis (Kwok *et al*, 1979). The interaction kinetics were concluded to be consistent with a loose noncovalent toxin-enzyme complex which could be reversed by phosphate (reaction product) before a covalent linkage occurred, but once the covalent linkage of the toxin to the enzyme formed (i.e. at low phosphate), the interaction (resulting in inhibition) was irreversible. This reasoning was used to explain why chlorotic symptoms of beans were relieved when plants were treated with phosphate immediately after inoculation (Kwok et al, 1979).

Further clarification of the mode of action of this toxin in relation to chlorosis, ornithine pool sizes, and OCTase inhibition was dependent on structural characterization of the toxin. Crude preparations used in early studies by Patil and coworkers were called phaseotoxin (Patil, 1974), one component of which was reported to be N-phosphoglutamic acid (NPG) (Patil *et al*, 1976). However, two groups independently demonstrated that synthetic NPG did not cause chlorosis or ornithine accumulation nor did it inhibit OCTase *in vitro* (Mitchell, 1979; Smith and Ruberg, 1979). Mitchell (1976) reported the active component produced in culture to be a tripeptide linked to sulfamyl phosphate; (N^{δ}-phosphosulfamyl) ornithylalanylhomoarginine which was given the trivial name phaseolotoxin. Corraborative structure studies now indicate that the principal toxin produced by pv. *phaseolicola* in culture and which possesses the required biological properties is phaseolotoxin (Macko this volume).

In a fashion analogous to the modification of tabtoxin to form tabtoxinine-β-lactam *in planta*, phaseolotoxin also appears to be converted to secondary toxic forms. When applied to bean leaves radioactive phaseolotoxin was rapidly converted to N^{δ}-phosphosulfamyl ornithine (PSorn) by sequential removal of homoarginine and alanine (Mitchell and Bieliski, 1977). Very little phaseolotoxin was found in infected bean leaves but the amount of PSorn found in the tissue was enough to account for the observed chlorosis. In contrast to the earlier failure to observe ornithine accumulation with impure phaseotoxin, a 10 ng dose of phaseolotoxin (presumably converted to PSorn) induced an 8,000 ng increase in ornithine per g fresh weight (Mitchell and Bieleski, 1977).

Direct comparison of phaseolotoxin and phaseotoxin A (NPG) for induction of chlorosis and ornithine accumulation revealed that the former caused both at 0.07 nmol while the latter caused neither at 34 nmol (Mitchell, 1979). At the same time purified phaseolotoxin inhibitied a commercial OCTase preparation from *Streptococcus faecalis* at a 1:16,000 inhibitor: substrate ratio whereas NPG did not inhibit the same preparation even at a 1:5 ratio. Smith and Rubery (1979) reported that NPG did not cause chlorosis in bean leaves, inhibit the growth of *Chlamydomonas reinhardi*, nor inhibit OCTase from bean leaves or *S. faecalis*. In contrast, pv. *phaseolicola* culture filtrates were active in all the above processes. Phaseolotoxin appears to be a generally potent inhibitor of ACTase from bacteria, fungi, plants, and animals with the notable exception of toxin producing strains of pv. *phaseolicola*. When grown at 18 C, the optimum temperature for toxin production, pv. *phaseolicola* produces an OCTase which is resistant to inhibition by phaseolotoxin (Straskawicz *et al*, 1980). Similar relationships appear to hold also for PSorn.

While OCTase inhibition and ornithine accumulation appear to be attributes of phaseolotoxin as well as the di- and mono-peptide analogs (Mitchell *et al*, 1981) the relationship of OCTase inhibition to chlorosis is less clear. Light is required for both chlorosis and ornithine accumulation but ammonia accumulation was not detected as with tabtoxin (Durbin and Uchytil, 1979) so an equivalent mechanism for inhibition of photophosphorylation appears unlikely. However, phaseolotoxin appeared to affect cross-pathway regulation in carrot cells where increases in carbomyl phosphate accompanied OCTase inhibition in these cells which in turn caused changes in pyrimidine biosynthesis (Jacques and Sung, 1978).

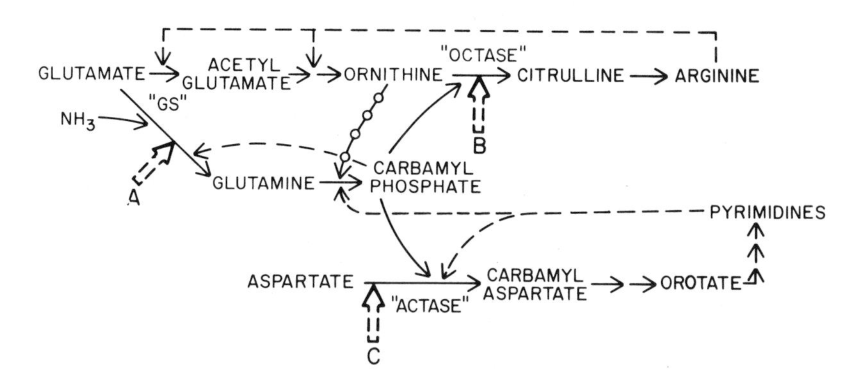

FIGURE 4. Biosynthesis pathways linking glutamine, arginine, pyrimidine synthesis. Dashed arrows (⇨) indicate proposed sites for inhibition of glutamine synthetase ("GS") by tabtoxinine-β-lactam (A), Ornithine carbomoyl tranferase ("OCTase") by phaseolotoxin (B), and aspartate carbamoyl transferase ("ATCase") by AAL-toxins (C). Reported feedback inhibition loops for various pathway metabolites are indicated by dashed lines (— — —) feedback activation loop involving ornithine and CAP synthetase is indicated by a dashed line with circles (o-o-o).

Carbamyl phosphate (CAP), synthesized by carbomoyl phosphate synthetase (CAP synthetase, E.C. 6.3.4.16), is a precursor for both arginine and pyrimidine biosynthesis. Following biosynthetic dogma, the pool size of CAP would be expected to be under tight regulatory control presumably at the site of its synthesis. (Fig. 4). CAP synthetase is feedback regulated by UMP while ornithine partially reversed the inhibition of the pea enzyme by UMP (O'Neal and Naylor, 1976). Accumulation of ornithine caused by OCTase inhibition thus would allow continued synthesis of CAP, albiet at a reduced rate, in the presence of elevated UMP levels with a resulting increase in pyrimidine synthesis via aspartate carbamoyl transferase (ACTase E.C. 2.1.3.3).

As noted by Kosuge and Kimple (1982), it may be significant that halo-blight bean leaves also contain elevated levels of methionine, lysine, β-alanine, and histidine which are metabolically unrelated to ornithine. The first three amino acids are derived from aspartate, a second precursor of pyrimidines, while the pyrimidines share a common precursor, 5-phosphoribosyl-1-pryophosphate, with histidine

(Miflin and Lea, 1980). If all these amino acids accumulate due to inhibition of OCTase, a broad alteration in metabolism which is normally metabolically interlocked could occur (Kosuge and Kimple, 1982). Increases in carbamyl phosphate, a reported inhibitor of glutamine synthetase (O'Neal and Joy, 1975), could lead to localized ammonia accumulation with attendant uncoupling of photophosphorylation in a fashion analogous to the previously stated tabtoxinine β-lactam situation. Alternatively, the consequence of reduced citrulline and arginine cannot be overlooked as contributing factors to chlorosis (Durbin, 1982).

A logical explanation for the observation that ornithine accumulation does not occur when citrulline and arginine are fed to phaseolotoxin treated tissue follows from the fact that arginine feedback inhibits the first and third enzymes committed to ornithine synthesis in higher plants, namely L-glutamate N-acetyltransferase (E.C. 2.3.1.1) and glutamate 5-phosphotransferase (E.C. 2.7.2.8) (Thompson, 1980).

E. AAL-Toxins

The causal agent of the stem canker disease of tomato, *Alternaria alternata* f. sp. *lycopersici*, (Grogan *et al*, 1975) produces phytotoxic metabolites in culture which appear to be essential determinants of pathogenicity. The toxins are not produced in culture by non-pathogenic *A. alternata*; they reproduce foliar interveinal necrosis; they are strikingly toxic only to tomato; and they are differentially toxic to certain genotypes of tomato (Gilchrist and Grogan, 1976).

Genetic studies of the host response to both the pathogen and crude toxin fractions indicated control by a single gene that segregates as a complete dominant for resistance to the pathogen but expresses incomplete dominance for tolerance to the toxin. The F_1 heterozygote (Rr) is resistant to the pathogen but is intermediate in toxin sensitivity to that of either parent, being 50 times more sensitive to the toxin than the resistant parent (RR) and 20 times less sensitive than the susceptible parent (rr) (Gilchrist and Grogan, 1976). Subsequent analysis of more than 2000 F_2 and 300 F_3 plants was in agreement with this inheritance pattern for a single disease-toxin locus (B. McFarland and D. Gilchrist, unpublished). The toxins produced by *A. alternata* f. sp. *lycopersici* thus conform to the host-specific class of pathotoxins (Daly, 1981).

Efforts were concentrated next on the purification and characterization of the toxin activity from culture filtrates. Two chromatographically separable components of the total toxin activity designated TA and TB were given the collective trivial name, AAL-toxin (Daly, 1981). In culture, TA and TB occur in a ratio of 3TA:1TB, each is host specific at the genotype level in tomato, and both have equivalent specific activity ~10 mg ml^{-1}. The major fraction, TA, has been shown to consist of two esters of 1,2,3-propanetricarboxylic acid (tricarballyate) and a novel amino pentol (1-amino-11,15-dimethylheptadeca-2,4,5,13,15-pentol) (Bottini and Gilchrist, 1981; Bottini et al, 1981).

In contrast to nearly all other host-specific toxins, ion leakage was not detected prior to the onset of visible necrosis (McFarland and Gilchrist, 1980). Preliminary spectroscopic data indicated the presence of an L-aspartate-like moiety in the AAL toxin structure. Although it was determined later that this moiety was tricarballylate, the suggestive presence of aspartate prompted initial mode of action studies in RR and rr tomato. L-aspartate and other compounds structurally related to aspartate were added to toxin preparations prior to biosassay to assess the temporal protection against toxin-induced necrosis. Interestingly, L-aspartate at 34 mM concentration (but not D-aspartate) reduced the toxin dilution end-point as much as 100-fold. The protective effect of L-aspartate was observed at concentrations as low as 34 μM in the standard leaf bioassay. Oxalacetate, threonine, and methionine of the aspartate family also protected at 34 mM, whereas amino acids in the glutamate, serine, and aromatic family were without effect at 34 mM. Pretreatment of rr tissue with L-aspartate at 34 μM also afforded protection against toxin-induced necrosis. Taken together, these results suggested that the toxin may act as an antimetabolite in some key metabolic function involving L-aspartate (McFarland and Gilchrist, 1980; McFarland and Gilchrist, unpublished).

Additional protection studies were undertaken to more closely focus previous results indicating that the site of toxin action was linked to either aspartate uptake (transport) or metabolism. In addition to the amino acids and keto acids previously mentioned, a total of 43 structurally or metabolically related compounds were tested as potential temporal protectants. Of all compounds tested the most striking protectant was orotate (McFarland and Gilchrist, unpublished) an intermediate unique to the orotic acid pathway of pyrimidine biosynthesis (Ross, 1981). Orotate was at least 10 fold more effective than L-aspartate, protecting at 3 μM. This suggested that a possible site of toxin action was the step catalyzed by aspartate carbamoyltransferase (ACTase, E.C. 2.1.2.3), an allosteric enzyme reported from several etiolated plant sources although it had not been studied previously in tomato.

ACTase catalyzes the condensation reaction between carbamyl phosphate (CAP) and aspartate (ASP) leading to the formation of carbamyl aspartate (CASP) (Fig. 4) which is the first committed step of pyrimidine biosynthesis in higher plants (Lovatt et al, 1979). Higher plant ACTase from etiolated lettuce seedlings (Neuman and Jones, 1962), etiolated mung bean seedlings (Ong and Jackson, 1972) and wheat germ (Yon, 1971a and 1971b) is regulated in vitro by the pathway end product UMP suggesting a regulatory function for this enzyme in higher plants (Ross, 1981).

Preliminary assay of ACTase activity from RR and rr tomato and mung bean (a non-host) by several methods indicated that the enzyme was inhibited by AAL toxin with the sensitivity order rr tomato ⩾ RR tomato ⩾ mung bean (McFarland and Gilchrist, 1980 and unpublished). However, the crude enzyme preparation from both tomato genotypes was extremely labile which precluded detailed kinetic characterization of the inhibition pattern. Thus, it was necessary to attempt to purify the enzyme, develop conditions for stabilizing the regulatory properties of the enzyme, and resolve questions regarding a suitable assay technique for the enzyme from tomato.

Convincing biological arguments for toxin induced necrosis using in vitro enzyme inhibition data depend on experimental conditions which permit the studies to be conducted with enzyme preparations that are both stable and responsive to known in vivo allosteric effectors throughout the course of the experiments. Also, one must assume that the enzyme encounters the toxin in the presence of the allosteric effectors in planta. Therefore, cell free preparations used for enzyme:toxin inhibition studies must be regulated in vitro and also must be assessed in the presence of the effectors.

Purification of ACTase from green tomato tissue was conducted recently in our laboratory (Gilchrist and McFarland, 1982) on cultivars of tomato representing RR and rr genotypes (with respect to AAL-toxin sensitivity) and etiolated mung bean as an internal check on published procedures. Five ACTase assay procedures were evaluated with the most sensitive and reliable procedure being method 2 of Prescott and Jones (1969) which is based on a colorimetric determination of carbamyl aspartate (CAA), the reaction product. The enzyme was purified 110 fold with 29% recovery of the crude extract activity. The isolated enzyme, which exhibited no metal cofactor or sufhydryl group requirement, had an estimated molecular weight of 79,000, pH optimum of 9.9, and was extremely labile with detectable loss in activity at 4 C in 4 hr and less than 20% remaining at 24 hr.

Extensive evaluation of potential stabilizing amendments revealed that 1 mM UMP stabilized the enzyme during purification. Although UMP is a natural feedback inhibitor of ACTase, a concentration of 1 mM did not inhibit the detection of activity upon dilution to 0.1 mM in the reaction mixture.

The partially purified ACTase from tomato displayed sigmoid kinetics with both substrates, carbamyl phosphate (CAP) and aspartate (ASP) with Hill values of 1.22 and 2.58, respectively. An ordered bi-bi mechanism for substrate addition with CAP the first ordered substrate was suggested by these results (Segel, 1975). Succinate, a dicarboxylic acid substrate analog, inhibited tomato ACTase competitively with respect to ASP (Kiapp = 32 mM) and uncompetitively with respect to CAP (I0.5 = 167 mM). Based on analyses of Dixon plots (Segel, 1975), UMP appeared to inhibit ACTase competitively with respect to CAP (Kiapp = 0.33 mM) and uncompetitively with respect to ASP (I0,5 ≃ 2.4 mM), the latter indicating UMP only binds to the enzyme complex after the binding of ASP. No evidence for a succinate:UMP synergism was observed when the mixed inhibitor data was analyzed by the method of Chou and Talaley (1977). ACTase from both RR and rr tissue was indistinguishable for all parameters, indicating a conservation of catalytic and regulatory properties in the absence of AAL-toxins.

ACTase from both genotypes was inhibited by AAL-toxins (TA + TB) in the absence of UMP, with the apparent dissociation constants (Kiapp) determined to be 2.4 and 4.2 mM for rr and RR respectively (McFarland and Gilchrist, 1982; McFarland and Gilchrist, unpublished). Assay of ACTase at 2.0 mM AAL-toxin decreased the Hill value for CAP to 0.72 (RR) and 0.64 (rr). Dixon plots (1/V vs. I) (Segal, 1975) were linear for CAP (2 to 16 mM) at 0, 0.5, 1.0, and 2.0 mM toxin and indicated uncompetitive binding, which presumably occurs subsequent to CAP binding. Hyperbolic velocity and linear Dixon plots suggested either single, or multiple noncooperative, AAL-site(s) on the enzyme. Comparable experiments with variable ASP at saturating CAP (10 mM) revealed noncompetitive toxin binding with respect to ASP for both genotypes. This type of binding differed from succinate.

The noncompetitive pattern of toxin vs ASP is taken to indicate that the enzyme binds AAL-toxin regardless of whether ASP is present and the substrate saturation alone will not relieve the inhibition (Segel, 1975).

In the presence of UMP, which itself decreases carbamyl phosphate binding (competitive inhibition), AAL-toxin interferes with the binding of carbamyl phosphate in a competitive manner for ACTase from both rr and RR tomato genotypes. In addition, an interesting synergism was detected to occur between toxin and UMP for the rr ACTase which was twice as great as that for the ACTase from tolerant tissue at ASP saturation. Furthermore, when assayed at variable ASP concentration the AAL-toxins appeared to be noncompetitive inhibitors with respect to aspartate, and in the presence of UMP (uncompetitive with respect to aspartate) act synergistically to produce a double, dead-end inhibitor complex with the net result that ACTase activity is decreased to zero by concentrations of UMP that are not normally (i.e., in the absence of the AAL-toxins) inhibitory. Additionally, this synergism occurred in a genotype-selective manner; the synergism was forty-fold greater for the rr genotype at aspartate saturation (20 mM). This difference roughly the same as that observed in AAL-toxin Kiapp values for the ACTase from the two genotypes.

The synergism between UMP and the AAL-toxins, if functional *in vivo*, would lead to a decreased sensitivity of the enzyme to the natural substrates and an increased sensitivity to the natural allosteric effector, UMP, as evidenced by the aspartate dependent decrease (and therefore increase in inhibition) in the Kiapp for UMP in the presence of the AAL-toxins.

Another factor, as yet unresolved, concerns the concentration of AAL-toxins required to demonstrate *in vitro* effects (1 to 5 mM). Although the concentration is consistent with *in vitro* substrate [app(S)0.5 for carbamyl phosphate = 2.16 mM and for aspartate = 3.14 mM] and effector [appKi-UMP for carbamyl phosphate = 0.33 mM and for aspartate = 2.4 mM] concentrations, and one to two orders of magnitude smaller than the app(K)i values for succinate [appKi-succ for carbamyl phosphate - 160 mM and for aspartate = 30 to 34 mM], it is still almost two orders of magnitude larger than the apparent concentration required to elicit necrosis in the sensitive tissues (0.02 μM) (McFarland and Gilchrist, unpublished). Both tentoxin (Steele et al, 1976) and phaseolotoxin (Ferguson and Johnston, 1980) were found to interact slowly and in a concentration dependent manner with their respective enzymes. Preincubation of those toxins with their enzyme in the absence of substrates resulted in up to a 100-fold increase in sensitivity to the toxins. Even though a 30 min preincubation time on ice was used for AAL-toxin, the 10 min prewarming period was always done in the presence of carbamyl phosphate thus is possible that a different incubation procedure might result in increased sensitivity to the AAL-toxins. The present data indicate the ACTase from tomato has a binding site for the toxin but exactly what the cumulative effect of ATCase inhibition might be in terms of altered pathway regulation is not known. The ultimate goal is to determine the sequence of events leading from toxin binding to ACTase and cell stress which is visualized, ultimately, as cell death.

VI. Conclusions

Elucidation of the molecular mode of action of pathotoxins has been an elusive goal even in cases where alterations are observed in specific physiological processes. The greatest obstacle appears to be detection of the chemically specific site(s) of pathotoxin interaction in the susceptible host cell. The greatest limitation may be the lack of chemically pure and structurally characterized toxin preparations. Attention to the molecular aspects of altered physiology suggests that there are a number of key mechanisms regulating intermediary metabolism of plants which should be considered as potential sites where pathotoxins can interact. Since enzymes are the cornerstones of metabolism their possible role in toxin induced disease physiology seems great.

This discussion focused on several selected examples where pathotoxins appear to interact directly with specific enzymes and emphasized the experimental strategies used to detect these interactions. The intent is not to imply that all molecular modes of action involve direct binding of pathotoxins to one or more specific enzymes. However, when one considers the biochemical factors regulating cellular integrity in plants it seems reasonable that one or a few specific enzymes will be affected initially, and directly, as cardinal determinents of the toxin induced metabolic dysfunction that leads to cellular stress. Understanding the location and role of such determinative enzymatic processes appears central to the resolution of molecular mode of action questions and, ultimately, the role of pathotoxins in disease.

Acknowledgments

Appreciation is expressed to Ann Martensen for the line drawings, Valinda Stagner for typing the manuscript, and Leona Gilchrist for editoral assistance.

References

Adams, D., and Yang, S. (1979). *Proc. Natl. Acad. Sci. U.S.A. 76,* 170.

Anderson, L., Nehrlich, S., and Champigny, M. (1978). *Plant Physiol. 53,* 835.

Arntzen, C. (1972). *Biochim. Biophys. Acta. 283,* 539.

Arntzen, C., Haugh, M., and Bobick, S. (1973). *Plant Physiol. 52,* 569.

Ballio, A. (1981). In "Toxins in Plant Disease" (R. Durbin ed.), p. 395 Academic Press, New York.

Ballio, A., Chain, E., DeLeo, P., Erlanger, B., Mauri, M., and Tonolo, A. (1964). *Nature (London) 203,* 297.

Beevers, H. (1969). *Ann. N. Y. Acad. Sci. 168,* 313.

Beffagna, N., Cocucci, S., and Marre, E. (1977). *Plant Sci. Lett. 8,* 91.

Bennet, J. (1975). *Phytochemistry 15,* 263.

Bhullar, B., Daly, J., and Rehfield, D. (1975). *Plant Physiol. 56,* 1.

Binder, A., Jagendorf, A., and Ngo, E. (1978). *J. Biol. Chem. 253,* 3094.

Boller, T., Herner, R., and Kende, H. (1979). *Planta 145,* 293.

Bonugle, K., and Davies, D. (1977). *Planta 133,* 281.

Bottini, A., Bowen, J., and Gilchrist, D. (1981). *Tetrahedron Lett. 22,* 2723.

Braun, A. (1950). *Natl. Acad. Sci. Proc. 36,* 423.

Braun, A. (1955). *Phytopathology 45,* 659.

Briggs, D. (1973). In "Biosynthesis and Its Control in Plants" (B. Milborrow, ed.), p. 219. Academic Press, New York.

Buchanan, B. (1980). *Ann. Rev. Plant Physiol. 31,* 341.

Chou, T., and Talaley, P. (1977). *J. Biol. Chem. 252,* 6438.

Cleland, W. (1967). *Ann. Rev. Biochem. 361,* 77.

Cohen, G. (1968). "The Regulation of Cell Metabolism." Holt, Rinehart, and Winston, Inc., New York.

Crosthwaite, L., and Sheen, S. (1979). *Phytopathology 69,* 376.

Culimore, J., and Sims, A. (1980). *Planta 150,* 392.
Daly, J. (1981). In "Toxins in Plant Disease" (R. Durbin, ed.), p. 331. Academic Press, New York.
Daly, J., and Barna, B. (1980). *Plant Physiol. 66,* 580.
Daly, J., and Knoche, H. (1982). In "Advances on Plant Pathology" (D. Ingram and P. Williams, eds.), Vol. 1, p. 83. Academic Press, New York.
Daly, J., and Uritani, I. (1979). "Recognition and Specificity in Plant Host-parasite Interactions". Univ. Park Press, Baltimore.
Davies, D. (1979). *Ann. Rev. Plant Physiol. 30,* 131.
Davies, D. (1980). In "The Biochemistry of Plants: A Comprehensive Treatise." (P. Stumph and E. Conn, eds-in-chief). Vol. 2 (D. Davies, ed.) p. 581. Academic Press, New York.
Day, P. (1979). "Genetics of the Host-Parasite Interaction". W. H. Freeman and Co., San Francisco.
Dennis, D., and Coultate, T. (1967). *Biochim. Biophys. Acta 146,* 129.
Dunkle, L., and Wolpert, T. (1981). *Physiol. Plant Pathol. 18,* 315.
Durbin, R. (1982). In "Phytopathogenic Prokaryotes" (M. Mount and G. Lacy, eds.), Vol. 1, p. 423. Academic Press, New York.
Durbin, R., and Steele, J. (1979). In "Recognition and Specificity of Plant Host-Parasite Interactions" (J. Daly and I. Uritani, eds.) p. 113. Univ. Park Press, Baltimore.
Durbin, R., and Uchytil, T. (1979). *Phytopathol. Mediterr. 18,* 199.
Durbin, R., Uchytil, T., Steele, J., and Ribeiro, R. (1978). *Phytochem. 17,* 147.
Ferguson, A., and Johnston, J. (1980). *Physiol. Plant Pathol. 16,* 269.
Filner, P., Wray, J., and Varner, J. (1969). *Science 165,* 358.
Fischer, R. (1968). *Science 160,* 784.
Flavin, M. (1975). In "Metabolic Pathways" (D. M. Greenberg, ed.), Vol. VII, p. 457. Academic Press, New York.
Frantz, T., Peterson, D., and Durbin, R. (1982). *Plant Physiol. 69,* 345.
Fulton, N., Bollenbacher, K., and Moore, B. (1960). *Phytopathology 60,* 575 (abstr.).
Fulton, N., Bollenbacher, K., and Templeton, G. (1965). *Phytopathology 55,* 49.
Gauman, E. (1954). *Endeavor 13,* 198.

Gilchrist, D., and Grogan, R. (1976). *Phytopathology* *66*, 165.
Gilchrist, D., and Kosuge, T. (1974). *Arch. Biochem. Biophys. 164*, 95.
Gilchrist, D., and Kosuge, T. (1975). *Arch. Biochem. Biophys. 171*, 36.
Gilchrist, D., and Kosuge, T. (1980). In "The Biochemistry of Plants: A Comprehensive Treatise" (P. Stumpf and E. Conn, eds-in-chief), Vol. 5, (B. Miflin, ed.), p. 507. Academic Press, New York.
Gilchrist, D., and McFarland, B. (1982). *Plant Physiol. 69*, 20 (abstr.).
Gilchrist, D., Woodin, T., Johnson, M., and Kosuge, T. (1972). *Plant Physiol. 49*, 52.
Giovanelli, J., Owens, L., and Mudd, S. (1971). *Biochim. Biophys. Acta 227*, 671.
Giovanelli, J., Owens, L. and Mudd, S. (1973). *Plant Physiol. 51*, 492.
Giovanelli, J., Mudd, S., and Datko, A. (1980). In "The Biochemistry of Plants: A Comprehensive Treatise". (P. Stumpf and E. Conn, eds-in-chief). Vol. 5 (B. Miflin, ed.) p. 453. Academic Press, New York.
Gnanamanickam, S., and Patil, S. (1976). *Phytopathology 66*, 290.
Grogan, R., Kimble, K., and Misaghi, I. (1975). *Phytopathology 65*, 880.
Guiz, C., Hirel, B., Shedlofsky, G., and Gadal, P. (1979). *Plant Sci. Lett. 15*, 271.
Halloin, J., DeZoeten, G., Gaard, G., and Walker, J. (1970). *Plant Physiol. 45*, 310.
Haschke, H., and Luttge, U. (1975). *Plant Physiol. 56*, 696.
Haschke, H., and Luttge, U. (1977). *Plant Sci. Lett. 8*, 53.
Heber, U., and Santarius, K. (1965). *Biochim. Biophys. Acta 109*, 390.
Hwok, O., Ako, H., and Patil, S. (1979). *Biochem. Biophys. Res. Comm. 89*, 1361.
Jacques, S., and Sung, Z. (1981). *Plant Physiol. 67*, 287.
Jensen, R. (1969). *J. Biol. Chem. 244*, 2816.
Johnson, H., Means, U., and Clark, F. (1959). *Nature 183*, 308.
Keys, A., Bird, I., Cornelius, M., Lea, P., Wallsgrove, R, and Miflin, B. (1978). *Nature (London) 275*, 741.
Koncewicz, M., Mathiaparanam, P., Uchytil, T., Sparapano, L., Tam, J., Rich, D., and Durbin, R. (1973). *Biochem. Biophys. Res. Comm. 53*, 653.

Kono, Y., and Daly, J. (1979). *Bioorg. Chem.* *8,* 391.
Kosuge, T., and Gilchrist, D. (1976). In "Encyclopedia of Plant Physiology" (R. Heitifuss, and P. Williams, eds.) Vol. 4, p. 679. Springer Verlag, Berlin.
Kosuge, T., and Kimple, J. (1982). In "Phytopathogenic Prokaryotes" (M. Mount and G. Lacy, eds.), Vol. 1, p. 365. Academic Press, New York.
Kuo, M., and Scheffer, R. (1970). *Phytopathology 60,* 1391.
Lamar, C., Sinden, S., and Durbin, R. (1969). *Biochem. Pharmacol. 18,* 521.
Lardy, H. (1980). *Pharmacol. Ther. 11,* 649.
Latzko, E., and Gibbs, M. (1969). *Plant Physiol. 44,* 396.
Lieberman, M. (1979). *Ann. Rev. Plant Physiol. 30,* 533.
Lovatt, C., Albert, L., and Tremblay, G. (1979). *Plant Physiol. 64,* 562.
McCarty, R. (1979). *Ann. Rev. Plant Physiol. 30,* 79.
McCarty, R., and Racker, E. (1966). *Brookhaven Symp. Biol. 19,* 202.
McFarland, B., and Gilchrist, D. (1982). *Plant Physiol. 69,* 21 (abstr.).
Magalhaes, A., Neyra, C., and Hageman, R. (1974). *Plant Physiol. 53,* 411.
Mankovitz, R., and Segal, H. (1969). *Biochemistry 8,* 3757.
Marre, E. (1979). *Ann. Rev. Plant Physiol. 30,* 273.
Meloche, H. (1967). *Biochemistry 6,* 2273.
Meyer, W., Kuyper, L., Lewis, R., Templeton, G., and Woodhead, S. (1974a). *Biochem. Biophys. Res. Commun. 56,* 234.
Meyer, W., Kuyper, L., Phelps, D., and Cordes, A. (1974b). *J. Chem. Soc. Chem. Commun. p. 339.*
Miflin, B., and Lea, P. (1976). *Phytochemistry 15,* 873.
Miflin, B., and Lea, P. (1980). In "The Biochemistry of Plants: A Comprehensive Treatise" (P. Stumpf and E. Conn, eds-in-chief). Vol. 5, (B. Miflin, ed.), p. 169. Academic Press, New York.
Mitchell, R. (1976). *Phytochemistry 15,* 1941.
Mitchell, R. (1979). *Physiol. Plant Pathol. 14,* 119.
Mitchell, R. (1981). In "Toxins in Plant Disease" (R. Durbin, ed.), p. 262. Academic Press, New York.
Mitchell, R., and Bieleski, R. (1977). *Plant Physiol. 60,* 723.
Neuman, J., and Jones, M. (1962). *Nature 195,* 709.

Nishimura, S., Kohomoto, K., and Otani, H. (1979). In "Recognition and Specificity in Plant-Host Parasite Interaction" (J. Daly and I. Uritani, eds.), p. 133. Univ. Park Press, Baltimore.
Oaks, A., and Bidwell, R. (1970). *Ann. Rev. Plant Physiol. 21,* 43.
O'Leary, M. (1982). *Ann. Rev. Plant Physiol. 33,* 297.
O'Neal, D., and Joy, K. (1973). *Nature New Biol. 246,* 61.
O'Neal, T., and Joy, K. (1975). *Plant Physiol. 55,* 968.
O'Neal, T., and Naylor, A. (1976). *Plant Physiol. 57,* 23.
Ong, B., and Jackson, J. (1972). *Biochem. J. 129,* 571.
Owens, L., and Wright, D. (1965a). *Plant Physiol. 40,* 927.
Owens, L., and Wright, D. (1965b). *Plant Physiol. 40,* 931.
Owens, L., Thompson, J., Pitcher, R., and Williams, T. (1968). *Biochim. Biophys. Acta 158,* 219.
Owens, L, Thompson, J., Pitcher, R., and Williams, T. (1972). *J. Chem. Soc. Chem. Commun. p. 174.*
Paulsen, J., and Lane, M. (1966). *Biochemistry 5,* 2350.
Patel, P., and Walker, J. (1963). *Phytopathology 53,* 522.
Patil, S. (1974). *Ann. Rev. Phytopathol. 12,* 259.
Patil, S., and Tam, L. (1972). *Plant Physiol. 49,* 803.
Patil, S., Kolattuky, P., and Dimond, A. (1970). *Plant Physiol. 46,* 752.
Patil, S., Youndblood, P., Christiansen, P., and Moore, E. (1976). *Biochem. Biophys. Res. Commun. 69,* 1019.
Platt, S., and Anthon, G. (1981). *Plant Physiol. 67,* 509.
Prescott, L., and Jones, M. (1969). *Anal. Biochem. 32,* 408.
Priess, J. (1982). *Ann. Rev. of Plant Physiol. 33,* 431.
Preiss, J., Biggs, M., and Greenberg, E. (1967). *J. Biol. Chem. 242,* 2292.
Racker, E., and Schroeder, E. (1958). *Arch. Biochem. Biophys. 74,* 326.
Raschke, K. (1975). *Ann. Rev. Plant Physiol. 26,* 39.
Rich, D., Mathiaparanam, P., Grant, J., and Bhatnagar, P. (1975). In "Peptides: Chemistry, Structure and Biology" (R. Walter, and J. Meienhof, eds.), p. 942. Ann Arbor Sciences. Ann Arbor, Mich.
Rich, D., Bhatnagar, P., Jasenski, R., Steele, J., Uchytil, T., and Durbin, R. (1978). *Bioorg. Chem. 7,* 207.
Ricard, J. (1980). In "The Biochemistry of Plants: A Comprehensive Treatise" (P. Stumpf and E. Conn, eds.-in-chief) Vol. 2. (D. Davis, ed.). P. 31. Academic Press, New York.

Ross, C. (1981). In "The Biochemistry of Plants: A Comprehensive Treatise" (P. Stumpf and E. Conn, eds.-in-chief). Vol. 6, (A. Marcus, ed.) p. 169. Academic Press, New York.
Rudolph, K., and Stahman, M. (1966). *Phytopathol. Z.* *57*, 29.
Scheffer, R., and Briggs, D. (1981). In "Toxins in Plant disease" (R. Durbin, ed.) p. 1. Academic Press, New York.
Scheffer, R., and Yoder, O. (1972). In "Phytotoxins in Plant Diseases." (R. Wood, A. Ballio, and A. Graniti, eds.) p. 251. Academic Press, New York.
Segel, I. (1975). "Enzyme Kinetics: Behavior and Analysis of Rapid Equilibrium and Steady-State Enzyme Systems." Wiley and Sons, New York.
Sellinger, O., and Weiler, P. (1963). *Biochem. Pharm. 12*, 989.
Shimke, R. (1969). In "Current Topics in Cellular Regulation" (B. Horecker and E. Stadtman, eds.), Vol. 1, p. 77. Academic Press, New York.
Shinke, R., and Mugebayashi, N. (1972). *Agric. Biol. Chem.* *36*, 378.
Sims, A., and Fergson, A. (1974). *J. Gen. Microbiol. 80*, 143.
Sinden, S., and Durbin, R. (1968). *Nature 219*, 379.
Sinden, S., and Durbin, R. (1970). *Phytopathology 60*, 429.
Smith, A., and Rubery, P. (1979). *Physiol. Plant Pathol.* *15*, 269.
Staskawicz, B., Panopoulos, N., and Hoogenraad, N. (1980). *J. Bacteriol. 142*, 720.
Steel, J., Uchytil, T., Durbin, R., Bhatnagar, P., and Rich, D. (1976). *Proc. Natl. Acad. Sci. U.S.A. 73*, 2245.
Steel, J., Uchytil, T., and Durbin, R. (1977). *Biochim. Biophys. Acta 459*, 347.
Steel, J., Durbin, R., Uchytil, T., and Rich, D. (1978). *Biochim. Biophys. Acta 501*, 72.
Stewart, W. (1971). *Nature (London) 229*, 174.
Stout, R., and Cleland, R. (1978). *Planta 139*, 43.
Taylor, P., Schnoes, H., and Durbin, R. (1972). *Biochim. Biophys. Acta 286*, 107.
Templeton, G. (1972). In "Microbial Toxins" (S. Kadis, A. Ciegler, and S. Ajl, eds.), Vol. VIII, p. 169. Academic Press, New York.
Ting, I., and Osmond, C. (1973). *Plant Physiol. 51*, 439.

Thompson, J. (1980). In "The Biochemistry of Plants: A Comprehensive Treatise." (P. Stumpf and E. Conn, eds.-in-chief), Vol. 5, (B. Miflin, ed.) p. 375. Academic Press, New York.
Turner, J. (1972). *Nature (London) 235,* 341.
Turner, J. (1981). *Physiol. Plant Pathol. 19,* 57.
Turner, J., and Debbage, J. (1982). *Physiol. Plant Pathol. 20,* 223.
Turner, J., and Turner, D. (1975). *Ann. Rev. Plant Physiol. 26,* 159.
Turner, J., and Turner, D. (1980). In "The Biochemistry of Plants: A Comprehensive Treatise." (P. Stumpf and E. Conn, eds.-in-chief) Vol. 2, (D. Davies, ed.) p. 279. Academic Press, New York.
Turner, N. (1973). *Am. J. Bot. 60,* 717.
Turner, N., and Granititi, A. (1969). *Nature (London) 223,* 1070.
Uchytil, T., and Durbin, R. (1980). *Experientia 36,* 301.
Wallsgrove, R., Lea, P., and Miflin, B. (1979). *Plant Physiol. 63,* 232.
Wheeler, H. (1978). In "Plant Disease: An Advanced Treatise." (J. Horsfall and E. Cowling, eds.), Vol. III, p. 327. Academic Press, New York.
Wheeler, H. (1981). In "Toxins in Plant Disease." (R. Durbin, ed.), p. 477. Academic Press, New York.
Winter, K. (1980). *Plant Physiol. 65,* 792.
Wolosuik, R., and Buchanan, R. (1976). *J. Biol. Chem. 251,* 6456.
Wolpert, T., and Dunkle, L. (1980). *Phytopathology 70,* 872.
Wooley, D., Pringle, R., and Braun, A. (1952). *J. Biol. Chem. 197,* 409.
Wooley, D., Schaffner, G., and Braun, A. (1955). *J. Biol. Chem. 215,* 485.
Yoder, O. (1980). *Ann. Rev. Phytopathol. 18,* 103.
Yoder, O. (1981). In "Toxins in Plant Disease." (R. Durbin, ed.), p. 45. Academic Press, New York.
Yon, R. (1971a). *Biochem. J. 121,* 18.
Yon, R. (1971b). *Biochem. J. 124,* 10.
Yon, R. (1973). *Biochem. Soc. Trans., Canterbury. 1,* 676.
Zeilke, H., and Filner, P. (1971). *J. Biol. Chem. 246,* 1772.

4

Roles of Toxins in Pathogenesis

SYOYO NISHIMURA AND KEISUKE KOHMOTO

I. Introduction

During the past two decades, the toxin concept in plant pathology has undergone great change. It generally had been believed that phytotoxic metabolites of plant pathogens were of limited importance in pathogenesis and lead only to visible injury on host plants, resulting in the reproduction of some of the disease symptoms, *e.g.*, chlorosis, wilting, necrosis or growth abnormalities. This disappointing outlook on toxin research has spanned nearly half a century, and began to disappear only when attention centered on a host-specific toxin in the early 1960s. It is now clear that host-specific toxins are important determinants in pathogenesis and can act as reliable surrogates for the pathogens that produce them. They thus seem to act as factors in induced susceptibility versus induced resistance (40). The development of toxin studies is now being watched with keenest interest. In the early investigations of host-specific toxins, however, the emphasis was mainly to elucidate their mode of action on susceptible plants. Toxin studies that focus attention only on mechanisms of physiological and biochemical action of toxins

TOXINS AND PLANT PATHOGENESIS
ISBN 0 12 200780 8

may not lead to a better understanding of the ecology of disease itself. The important things are to identify the precise site with a tissue or cell of a host where the critical interaction occurs, leading to the triggering of induced susceptibility.

The critical question of the fundamental role of toxin production by the pathogen that makes it capable of causing successful pathogenesis both in host tissues and in an agro-ecosystem has been left largely unstudied. However, some remarkable recent advances would indicate that this situation is undergoing some interesting changes. These include genetic analyses of toxin production in pathogens, the importance of the release of toxin from germinating spores of pathogens, the concept of toxin producers as mutants, and the concept of toxins as suppressors of general defense mechanisms in plants.

In this review, no attempt is made to present a complete survey of the existing literature, nor will any attempt to deal with the role of common phytotoxic metabolites of pathogens. Rather the aim is to concentrate for the most part on our information currently available on *Alternaria alternata* infections which are dependent on a host-specific toxin. The material here is divided into two sections. The first considers significance of toxin production in pathogenesis, and the second deals with toxins and evolution of parasitism. Previous reviews (14, 16) and chapters (17-19) have dealt with other aspects of the subjects. Research advances in this field also have been discussed by other reviewers (31, 38, 40).

II. Significance of Toxin Production in Pathogenesis

A. *Host-Specific Toxins and Pathogenicity*

The growing lists of host-specific toxins have been described many times in the past (14, 17-19, 30, 33, 40). Nearly fifteen examples are now known in the literature. A brief summary is sufficient for the purposes of this chapter. The present knowledge of such toxins has come almost entirely from so-called "saprophytic pathogens." The fungi, even if they lose toxin productivity, can be expected to survive as saprophytes. Included are familiar fungi, such as *Alternaria*, *Helminthosporium*, as well as others less well known. Strangely, most of these appeared to be as causes of so-called "man made-disease," which occur only on newly bred or introduced cultivars of crop plants; unfortunately, these cultivars are controlled by genetically susceptible genes.

There is a group of pathogenic *Alternaria* spp. that have been classified mainly on the basis of their host-specificities. Among these are *A. kikuchiana* affecting Japanese pear, *A. mali* affecting apple, *A. citri* affecting citrus, *A. alternata* f. sp. *lycopersici* affecting tomato, *A. alternata* affecting strawberry, and *A. longipes* affecting tobacco. Each of these fungi is now known to produce a host-specific toxin that appears to be responsible for its specific pathogenicity and host range (14, 19). For convenience, these *Alternaria* toxins have been given short-hand names (Table 1). None of these has a reported sexual stage but all are morphologically similar enough in their asexual stages to be classified as a collective species *Alternaria alternata* (Fries) Keissler.

Table 1. Alternaria alternata Toxins Known to Date (19)

Pathotype of A. alternata (previous name)	*Disease*	*Host*	*Toxin*
1. Apple pathotype (A. mali)	*Alternaria blotch of apple*	*Apple & Japanese pear*	*AM-toxin*
2. Citrus pathotype [a] *(A. citri)*	*Brown spot of citrus*	*Citrus*	*AC-toxin*
3. Japanese pear pathotype (A. kikuchiana)	*Black spot of Japanese pear*	*Japanese pear*	*AK-toxin*
4. Strawberry pathotype	*Alternaria black spot of strawberry*	*Strawberry & Japanese pear*	*AF-toxin*
5. Tobacco pathotype (A. longipes)	*Brown spot of tobacco*	*Tobacco*	*AT-toxin*
6. Tomato pathotype (A. alternata f.sp.lycopersici)	*Stem canker of tomato*	*Tomato*	*AL-toxin*

[a] *Two different isolates, pathogenic to Citrus reticulata (e.g., cv. Dancy tangerine) and pathogenic to C. jambhiri (rough lemon) were found. Each isolate produces selective toxin to each host plant.*

A. alternata is an ubiquitous fungus commonly isolated from various dead plant materials. It also belongs to a large diffuse species composed of many strains that vary in their ability to attack different plants as a weak pathogen, causing indefinite or opportunistic diseases; most of these isolates do not exhibit specific pathogenicity on their plants. Regardless of similarities on conidial morphology, the above six examples of *A. alternata* pathogens have a limited host range. On a chemical basis, the virulence and host-specifities can be a manifestation of the production of a distinctive host-specific toxin. Production of such a toxin appears to be

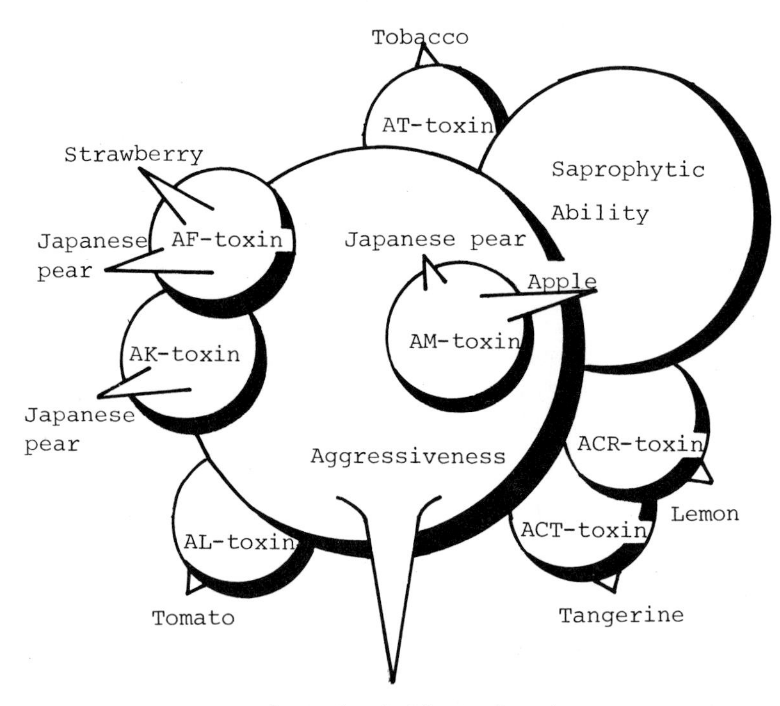

Figure 1. Diagrammatic presentation of constituent elements of pathogenicity in Alternaria alternata pathogens. The pathogenicity is composed of three parts, saprophytic ability, aggressiveness, and specific virulence. Small globes on the largest globe (indicating aggressiveness) represent virulence of each pathotype, exhibited by host-specific toxin and host range (thorns).

essential for a given *Alternaria* pathogen; no other significant differences to distinguish among these fungi have been established to date. Thus, we believe that each pathogen should reasonably be called a distinct "pathotype" of *A. alternata*, rather than defining them as different species (14, 17, 19, 22). Our initial model has been refined, and is diagrammed (Fig. 1) to show the nature of specific virulence or toxin production in *A. alternata* (19). It shows that the pathogenicity in broad sense is composed of three parts, saprophytic ability, aggressiveness (22) (the ability to penetrate plant tissues or cellophane without causing disease), and specific virulence involving a host-specific toxin. The conceptual model, although it may be applied at the moment to the relationship among several pathotypes of *A. alternata*, may prove to be incorrect in detail. In its support, nevertheless, the results of investigations of toxin producers in the laboratory and in the field summarized in the following sections offers additional evidence.

B. Toxins as Initiation Factors of Successful Pathogenesis

Much research was begun on the host-specific toxins produced by plant pathogens during the 1960s and 1970s. However, most of the research was biased toward toxin isolation and purification, or the mode of biochemical action on susceptible plant tissues, because of their interesting and unique host selective phytotoxicity when compared with general fungal phytotoxins. Since the late 1970s several groups of chemists, in co-operation with plant pathologists, began to elucidate their chemical structures. Accumulated knowledge of these research fields helped undoubtedly to establish the importance of toxins in plant pathology. However, it would appear that an understanding of the exact roles of such toxins in relation to successful pathogenesis as well as to the epidemiology of disease is the critical need. This may be the ultimate objective of toxin study in plant pathology.

Interestingly, all known host-specific *Alternaria* toxins can be detected from the spore-germination fluids of each virulent pathogen, but not from those of avirulent ones (Fig. 2, A & B). This is a significant point, because such productivity seems to be characteristically associated with host-specific toxin, and suggests the very early participation of toxin at the site of the initial contact of the inoculated pathogen and host surface. We (19) used the term "release toxin" to denote a host-specific toxin from germinating spores of pathogens. Pathologically, the findings emphasize the

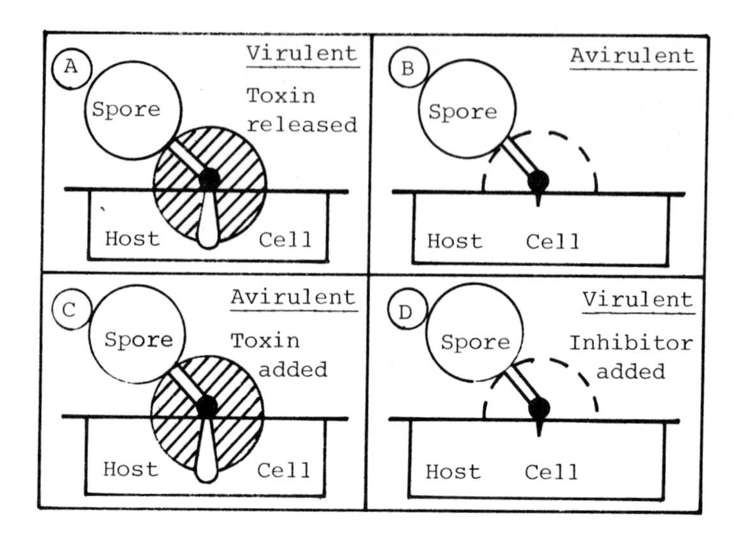

Figure 2. Diagram illustrating the importance of host-specific toxin production from germinating spores of pathogens.

role of these toxins as a primary determinant (32), virulence factor (40) or initiation factor (16).

The importance of release toxin was first proposed in 1965 by Nishimura and Scheffer (21), who observed that virulent spores of *Helminthosporium victoriae* germinated and released the host-specific HV-toxin, in 3 hr on glass slides or on oat leaves. The concept of release toxin helped to focus on the molecular interactions between host plant and parasite, and provided a framework for dealing with the positive evaluation of a role for toxin in pathogenesis. Yoder and Scheffer (41) demonstrated that when HV-toxin was added to infection droplets of avirulent mutant that had lost the ability to produce the toxin, the mutant spores then colonized susceptible but not resistant oat leaves (Fig. 2, C). Later, essentially, the same results were obtained in a study of many combinations, *e.g., H. carbonum* - HC-toxin - corn (2), *A. alternata* Japanese pear pathotype - AK-toxin - Japanese pear (28), and *A. alternata* tobacco pathotype - AT-toxin - tobacco (34). Paradoxically, data obtained only by the use of culture filtrates of pathogens can not provide such clear evidence for an important role of toxin in early events of pathogenesis. We no not know how widespread release toxins may be in nature, but it seems likely that we will be able to detect them from many other plant pathogens, particularly from so-called "saprophytic pathogens."

Wheeler (38) coined the term of "telepathogenesis" to denote "the ability to cause disease at a distance without being in physical contact with the plant." He pointed out that some, perhaps many, toxin-producing plant pathogens may be telepathogenic, and the toxins may act as symptom-development agents or else as predisposing agent in the absence of a living pathogen. This definition is somewhat analogous to the concept of release toxin. The following experiment with AK-toxin, produced by the *A. alternata* Japanese pear pathotype, will be concerned with the influence of toxins as telepathogenic factors in plant disease; consequently the emphasis is placed on the release toxin that predisposes plant toward greater susceptibility.

With host - pathogen combination of *A. alternata* Japanese pear pathotype and Japanese pear, the first physiologically detectable event induced by spore inoculation is an increased loss of electrolytes from susceptible leaves (28). A careful measurement of electrical conductivity reveals two different phases; at 2 - 6 hr and at 9 hr after inoculation. As mentioned previously, AK-toxin from the pathogen can be released during germination on leaves. It also causes an almost instantaneous increase in electrolyte loss from susceptible tissues, but not from resistant tissues (16, 18, 27, 28). At 4 hr after inoculation, one germinating spore yields approximately 10^{-6} μg toxin, which is capable of disturbing the metabolic activities of approximately 100 host cells. These data lead to the conclusion that first increase in electrolyte loss at 2 - 6 hr is caused by AK-toxin production from germinating spores prior to host invasion. The second increase at 9 hr after inoculation is probably due to the toxin from the host-invading hyphae. Visible symptoms become evident by 12 - 20 hr after inoculation and they appear as tiny, dark necrotic spots on susceptible leaves. When resistant leaves are inoculated with virulent spores, or when susceptible leaves are inoculated with avirulent spores, there is no increase in electrolyte loss. The above step-like increases in leakage add support to the hypothesis that pathogenic *A. alternata* induces disease via toxin production prior to penetration of host. A similar mode of host responses during pathogenesis can be shown in other combinations, apple pathotype of *A. alternata* and apple leaves via AM-toxin (8), and strawberry pathotype of *A. alternata* and strawberry leaves via AF-toxin (15).

C. Metabolic Inhibition of Toxin Production

A fundamental question remains however. How do the release toxins from germinating spores of pathogens trigger or elicite the pathologically important responses in susceptible host tissues ? Before turning our attention to this problem, recent contributions concerning the regulation of toxin production will be introduced. If the production of release toxin is inhibited by some means during spore germination without any inhibitory influence on other functions, then the virulence exhibited by the pathogen can conceptually be expected to be reduced (Fig. 2, D).

Observations with Japanese pear pathotype of *A. alternata* indicated that cerulenin (5, 26), an antibiotic from *Cephalosporium caerulens*, almost completely inhibits the initiation of AK-toxin production by germinating spores of virulent isolates (20). Cerulenin is known to be specific inhibitor of condensing enzyme activity among the enzymes involved in fatty acid biosynthesis (23, 24). When the spores are placed on susceptible pear leaves in the presence of cerulenin (5 - 50 μg/ml), the appearance of black necrotic spots obviously decreases; those spores germinate normally on the leaves, but the germ tubes do not produce AK-toxin. This implies that cerulenin inhibits the production of this toxin synthesized *in vivo* through fatty acid or related intermediates. In fact, the chemical structures of AK-toxin I and II are suggestive of the participation of fatty acid related moieties (13). That sulfure-containing amino acids, *e.g.*, ethionine, methionine and cysteine, can also be effective was recently shown by our group (Tsuge and Nishimura, unpublished).

In any case, evidence obtained with cerulenin demonstrates, for the first time, that toxin metabolism in germinating spores of pathogens can be regulated artificially by chemicals. We believe that other chemicals with similar functions should be more extensively investigated as tools for studying the exact role of toxins in pathogenesis. It is possible that such studies could be used to develop new types of disease protectants in the future. The idea of protecting plants against disease by means of inhibiting toxin production is not new, but the time to exploit it may now be maturing.

In addition to the above chemical treatment, physical treatment of spores produced interesting evidence for the importance of release toxin in pathogenesis. Heat shock (*e.g.*, in water at 50 C for 10 sec) of the dormant spores of AK-toxin producer of *A. alternata* before inoculation suppressed their virulence on pear leaves; spore germination is little inhibited, but the amount of toxin released is significantly inhibited

(36). Such heat-induced inactivation of toxin production is reversible and the spores recovered almost completely within a day after the treatment at room temperature. Although these results must be interpreted with caution, they appear to reflect a change in enzymes or more complex metabolic systems which are involved in toxin production in fungal cells.

D. Toxins as Suppressors of Defense Reaction

Disease resistance in plants may be only relative, and plants constantly face conditions of predisposition in nature. Yet, in spite of the fact that general concept of predisposition is nearly a century old, it is surprising how little work has been accomplished on its molecular basis. Some toxin researchers have had an interest in taking a closer look at how and why toxins affect the predisposition or general defense reaction in plants. It is appropriate to ask as to which morphological, biochemical or physiological changes in toxin-treated plant tissues are directly related to the successful colonization of pathogens.

Before discussing this problem, current concepts of parasitism or host - parasite specificity will be introduced briefly. In 1981, Bushnell and Rowell (1), and Heath (7) independently proposed a generalized concept, using a model. It is speculated that a fungal parasite that can cause disease in a given higher plant species has the ability to avoid or to negate the sequences of general defense, both preformed and induced. Such a parasite is basically compatible with its host and thus has established compatibility at the species level of specificity. They predict that specific and active accommodation to the parasite to its host includes the production of host-specific toxins or the secretion of fungal suppressors which prevent specific defense reaction from taking place. Ellingboe (3) proposed that the idea that specificity at the race - cultivar level is superimposed on the "basic compatibility" at the fungus species - plant species level. This was based on the genetically accepted gene-for-gene theory in which single corresponding genes in host and parasite condition incompatibility. Their statements cannot be regarded as isolated discussions for toxin researchers; we must be concerned particularly with their consideration about the basic compatibility.

Something must happened in susceptible tissues within a short time after toxin exposure. How important are these events for rendering defense factors or reactions in the tissues ? The experimental evidence that toxic metabolites of pathogens induce susceptible conditions of host plant was probably first advanced by Gnanamaikam and Patil (4) in 1977. They used the combination of *Pseudomonas syringae* pv. *phaseolicola* causing halo blight disease of bean and its halo-inducing toxin, phaseolotoxin. In addition to its roles as an inducer of chlorosis and as a specific inhibitor of ornithine carbamoyltransferase, phaseolotoxin may also be involved in the establishment of its bacterial pathogens in bean tissues. When plants of resistant bean cultivars are inoculated with the pathogen, they respond by producing hypersensitivity and phytoalexins. Bacterial growth is sharply curtailed, and no toxin is detected in the tissues of such plants. However, when plants of the same cultivars are treated with the toxin prior to inoculation, the production of hypersensitivity and phytoalexin is suppressed, bacterial multiplication increased, and typical susceptible symptoms are observed. The toxin-inducing suppression of defense reaction can be found to be specific against only pv. *phaseolicola* and not against several other bacteria.

In cases of host-specific toxins from fungal pathogens, regrettably no direct evidence for this interesting research field has been produced. The working hypothesis proposed by Scheffer and Yoder (33) in 1972, toxin-producing pathogens are essentially saprophytes incapable of invading tissues before they are killed or their activities altered, still remains only as a sobering thought.

Our still fragmentary knowledge concerning *A. alternata* toxins is summarized below with several lines of evidence, but nothing conclusive is available at this time.

(1) Studies on the altered sensitivity of host tissues against toxin-producing pathogens and its host-specific toxin under high temperature stress are of much interest. When Japanese pear leaves susceptible to black spot disease were subjected to a thermal shock (*e.g.*, in water at 30 C for 10 sec), they became more susceptible to the toxin-producing *A. alternata* (19, 27). They also became susceptible even to toxin-less mutant or saprophytic *A. alternata*. In resistant leaves, however, such a thermal treatment did not change the original insensitivity to the toxin. Instead, it caused a loss of resistance to the fungus; even saprophytic ones could invade the tissues, as recognized by the formation of black necrotic spots on the heat-shocked leaves. All of these altered properties were reversible, and the leaf tissues recovered their original states within 12 hr after the treatment. The above results can be interpreted as resulting from two different events, although these must be interpreted with caution. The toxin sensitive site (probably toxin receptor)

which exists only in susceptible tissues and the resistance-activating site in both susceptible and resistant tissues may be changed independently by heat shock; the possibilities would depend on the temporary changes of the structural nature of the sites by heat shock.

(2) We can ask then, what is the potential for the induction of general defense in plants, independently of their genetically determined susceptibility or resistance ? This problem is also of general interest in another connection with the pathological mode of action of the release toxin from germinating spores of pathogens. We will restrict our discussion mainly to our example of *A. alternata* toxins.

Germinating spores of *A. alternata* isolates, independently of their toxin productivity, produce nonphytotoxic metabolite(s)----so-called "inducer(s) or activator(s)"----which nonspecially induce resistance in host tissues against toxin-producing *A. alternata* (6). In Japanese pear leaves, an interval of 4 to 6 hr after treatment with the inducer was necessary for the induction of resistance to be expressed, resulting in the significant production against AK-toxin producers of *A. alternata*. When AK-toxin was added to the inducer solution, the protection effect completely disappeared. This means that the action of AK-toxin against susceptible pear leaves already was completed before the inducer function. Furthermore, the protective effect of inducer can be suppressed by pretreatment of leaves in water at 50 C for 20 sec, but not by the post-treatment of 4 to 6 hr later.

The data support the following conclusions: (a) pear tissues possess potential resistance mechanism to fungal invasion; (b) the resistance is induced by certain substance(s) (probably heat-labile glycoprotein) released from germinating spores of most of *A. alternata* isolates; and (c) in pathological terms AK-toxin action may play a role as a suppressor for resistance mechanisms in susceptible pear tissues.

Our initial model shows that the release toxin interacts with a hypothetical site in susceptible tissues and that the inducer from the pathogen activates nonspecially the induction of general defense in plants (Fig. 3). Phytoalexin-like compounds have not been detected in the leaves of most rosaceous plants, including pear and strawberry leaves. Instead, the efficacy of the inducer was detectable by the production of so-called "infection-inhibiting factor," although it may also be considered as a multicomponent system including physical (*e.g.*, formation of wall appositions) and chemical events. Chemical characterization on the infection-inhibiting factors from pear and strawberry leaves is being undertaken in Nishimura's and Kohmoto's laboratories.

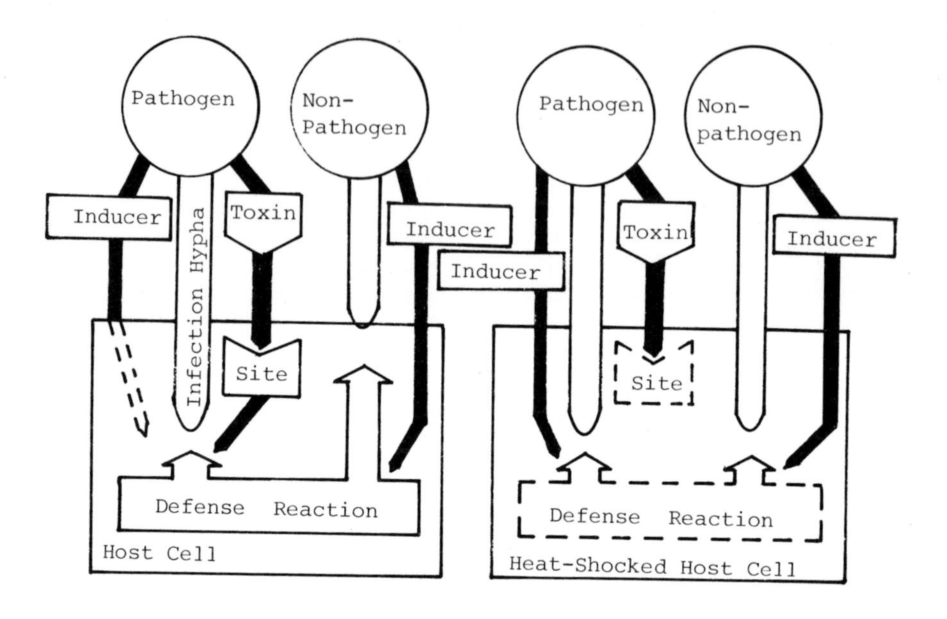

Figure 3. A model illustrating the relation among host-specific toxin from germinating spores of pathogen, its binding site in host and activator for induction of defense reaction in host.

III. Toxins and Evolution of Parasitism

A. Toxin Producers as Biochemical Mutants

As is often discussed for fungi that produce host-specific toxins, when the field isolates are kept in cultures, particularly at high temperatures (*e.g.*, at 35 C), most of these suddenly lose ability to produce toxin (17, 19, 29, 33). The mechanism is still far from understood. The reverse, from nonpathogenic type to pathogenic type (toxin producer) should also be true in laboratory conditions.

This prediction could be experimentally addressed (19). Young pear leaves, susceptible and immune to black spot disease, respectively, which were freshly harvested from the growing plants in a clean greenhouse, were sprayed with spore

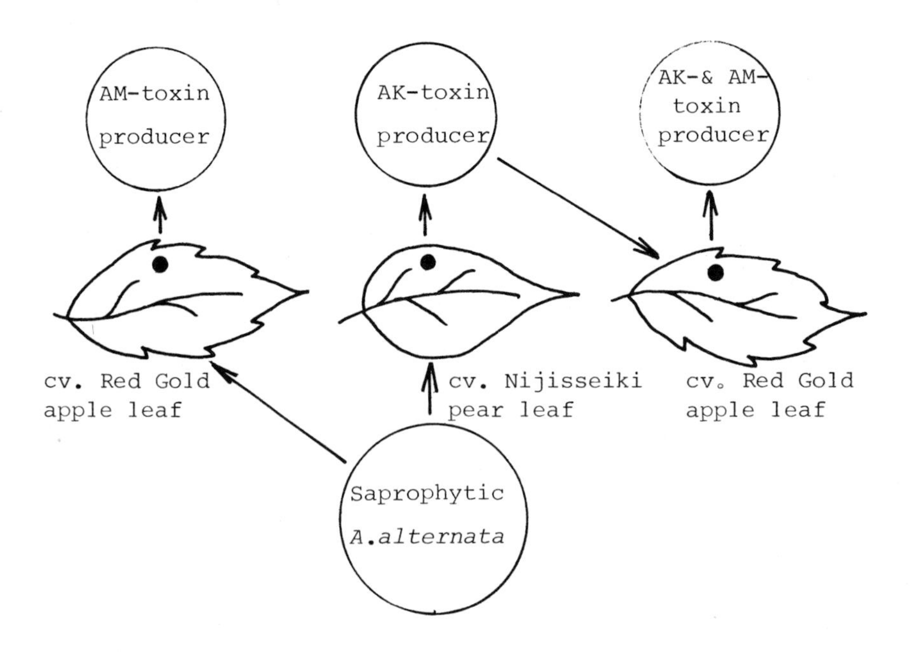

Figure 4. Diagram illustrating the leaf bioassay for detection of three different toxin-producing mutants in saprophytic Alternaria alternata.

suspensions of apparently saprophytic *A. alternata* and incubated in a moist chamber. Leaves sprayed with sterilized water served as a control. In two days, small necrotic black spots were observed very occasionally on the susceptible leaves only, but none on resistant (immune) ones or the controls ones. Some isolates obtained from the spots resembled the authentic pear pathotype of *A. alternata* already known in terms of pathogenicity and AK-toxin production (Fig. 4). As with the AK-toxin producer, AM-toxin producers could also be isolated from apple leaves susceptible to Alternaria blotch of apple and sprayed with the saprophytic spores of *A. alternata*. Furthermore, double toxin producers, that produced both AK- and AM-toxins in cultures and were equally pathogenic to Japanese pear and apple leaves, could be detected, when spores of AK-toxin producer were inoculated on susceptible apple leaves.

In the above results, we assume that our original inoculum must have contained previously existing toxin producers, and conceptually the host surface has contributed favorably to the detection of toxin mutants as a selection medium. It seems rather unlikely that there is any host effect or host factor involved with the expression of toxin mutants in the saprophytic *A. alternata* spores used. However, there is another report of the apparent mutation of pathogenicity, thereby causing toxin productivity to be acquired. Strobel's group (12, 29) indicated that when the attenuated cultures (toxinless mutants) of *Helminthosporium sacchari* were grown on material obtained from the water wash of sugarcane leaves susceptible to eye spot disease, the production of host-specific HS-toxin could resume. Two of the activators were identified as serinol (2-amino-1,3-propandiol) and a novel glycolipid. The leaves of a resistant clone did not possess such activators. Thus, they predicted that many pathogenic organisms could use key metabolites as compounds to regulate aspects of their pathogenicity, such as toxin production. This may be a chemical evidence for "bridging host theory" which certain pathogens change their virulence during passage through resistant or susceptible hosts, proposed by Ward's group long ago (37).

How could toxin producers of *A. alternata* mutate and lose toxin production ? Conversely, how could saprophytic *A. alternata* mutate to toxin producers ? We have great interest in this problem, because of its immense practical implications. Distinct pathotypes of *A. alternata*, whose specific virulence likely depends on the potential to produce toxin, may be called "biochemical mutants" pertaining to host-specific toxins, although the present study is not entirely conclusive.

Yoder (40) predicted with his own data and Leonard's work (9-11) that a new race (race T) of *Helminthosporium maydis* appeared, possibly from a mutation, a genetic recombination, or from rapid increase in an endemic form. Race T produces host-specific HMT-toxin and is highly virulent to corn cultivars with cytoplasm of the Texas (T) male-sterile type. Race O probably does not produce the toxin and is not highly virulent to T-cytoplasm corn. If the recent mutation hypothesis, that the current population of race T race arose recently after sufficient acreage of T-cytoplasm corn, is logical to predict genetically, mutation experiments from race O to race T may be appropriate, as with the cases of *Alternaria* toxins. The appearance of *Phyllosticta maydis* causing yellow leaf blight of T-cytoplasm corn could also be interpreted to support the recent mutation hypothesis (40).

In the following section, indirect evidence may be offered for natural mutation in agroecosystem.

B. Toxins as Agents of Selection Pressure

Toxin studies would help to clarify other important questions in plant pathology. How do the fungal infections depending on host-specific toxins and the related fungal metabolites break out drastically in the field ? What is the ecological basis for this high level occupation of such a pathogen in agroecosystem ? What is the evidence that toxins may act as a selection pressure under field conditions ?

We had previously investigated the seasonal changes in the number of airborne spores of *Alternaria alternata* in both cv. Nijisseiki (susceptible to black spot disease) and cv. Kosui (immune to the disease) Japanese pear orchards, by monitoring their AK-toxin-producing ability (35). Surprisingly, only about 2 % of the airborne *A. alternata* spores trapped from commercially managed Nijisseiki pear orchard, and none from Kosui pear orchard produced the toxin. The detection of toxin-producing spores centered on the two periods from May to June and from September to October; these periods just correspond to the development periods of young leaves that are highly sensitive to fungus and its toxin. A analogous phenomenon was reported by Odvody *et al.* (25) for *Periconia circinata* causing milo disease of sorghum, where only 13 % of the isolates from soil and 34 % from roots of susceptible sorghums in a milo disease nursery, and none from resistant sorghums produced host-specific PC-toxin.

We do not know whether abundantly trapped nontoxin-producing (nonpathogenic) *A. alternata* were toxin-less mutants or whether they were saprophytes essentially unrelated to toxin producers. An important fact, which is worthy of comment in our field survey, was that, in the immune Kosui pear orchard, the probability of trapping AK-toxin producers was not a zero, but was certainly possible even with an extremely low rate. If we persevere in taking the trapping survey of a large number of airborne spores, then we should be successful in the capture of the toxin producers. Surprisingly, other known toxin producers, *e.g.*, AC-toxin and AF-toxin producers, as well as AK-toxin producer, could be trapped; there was no cultivation of mandarin or tangerin and strawberry susceptible to brown spot of citrus and Alternaria black spot of strawberry, respectively, around the Kosui pear orchard surveyed. Thus, the above field survey is suggestive of the ceaseless emergence of natural mutation from apparently saprophytic *A. alternata* to several different toxin producers which are specific pathotypes, independently of the cultivations of susceptible genotypes. Mutation to toxin production probably comprises an extremely minute proportion of the field population of *A. alternata* before the introduction of susceptible plants. Monoculture of such a plant apparently increases the proportion of a specific pathotype very rapidly. If there is no chance to meet with uniformly genetically susceptible genotypes, the property of toxin production may not be displayed to any extent, and must be soon lost (Fig. 5).

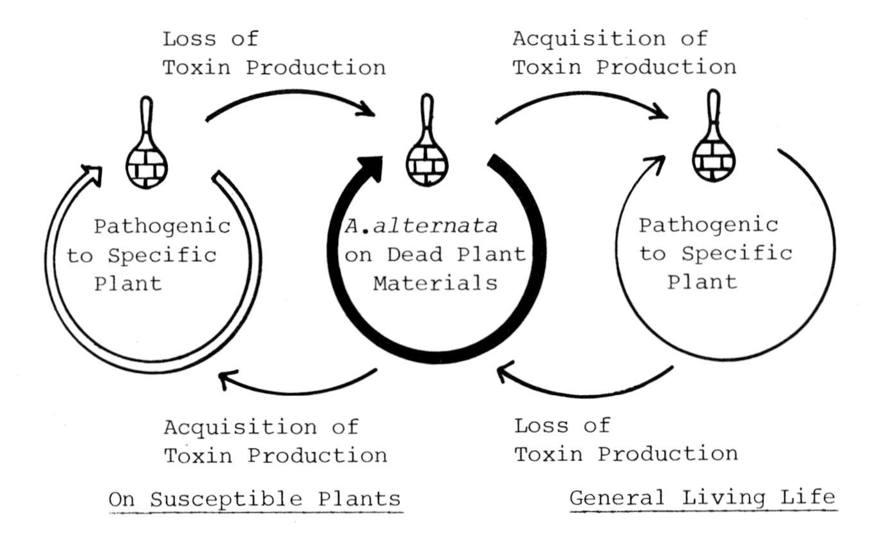

Figure 5. Diagram illustrating the living pattern of Alternaria alternata pathogens in field.

It is important to note, although it has no direct connection with toxin studies, that many practical pathologists have overlooked the quantification of pathogenic spores of airborne *A. alternata* in susceptible cv. Nijisseiki pear orchards. More precise information on only the pathogenic population during the growing seasons of the fruit trees has both basic ecological significance and practical implications for integrated control strategies of the disease. This is the only prospect for agricultural applications of our knowledge of toxins affecting plants.

To what extent can cultivation of susceptible genotypes influence the parasitic fitness of pathogens in natural ecosystems ? In other words, are there any effects of toxin on the increased disease efficacy of pathogens that can result in a rapid occupation in agroecosystems ? If so, how can such favorable association with toxin occur, and under what environmental conditions ? Unfortunately, the answer to these questions all remain open.

Historical developments of the catastrophic diseases involving toxins may give somewhat a hint, and are suggestive of at least three types of field emergence of diseases; (1) sudden emergence after widespread planting of susceptible genotype, *e.g.*, black spot of Japanese pear in Japan, Victoria blight of oat in U. S. A., leaf blight of corn in many countries, Alternaria black spot of strawberry in Japan, and stem canker of tomato in U. S. A. ; (2) sudden emergence after several decades of the cultivation of susceptible genotype, *e.g.*, Alternaria blotch of apple in Japan, and stem canker of tomato in Japan; and (3) universal emergence in many countries, *e.g.*, eye spot of sugarcane and brown spot of tobacco. For the details of those historical developments, the reader is referred to reviews by Pringle and Scheffer (30), Scheffer (31), Yoder (40), and Nishimura *et al.* (19). The reason for differences among those types of disease occurrence is not known at present, but a plausible explanation would appear to involve many factors, *e.g.*, genetic background (homozygous or heterozygous susceptible genotypes), mutational ability of the pathogen, and rapid changes in crop cultivation including fungicides applied.

There is another interesting problem to show that the competition for ecological occupation between two different pathotypes of *A. alternata* on the same susceptible host plant may occur in field. In the mid-1970s, a newly bred cultivar of strawberry, Morioka-16, was released in the northern district of Japan. A problem with a new disease, Alternaria black spot, caused by a new pathotype of *A. alternata* was almost immediate; the associated variation in the fungus---- the change from saprophyte to virulence on this cultivar---- might have occured as a recent mutation. The isolates were highly pathogenic only to this cultivar, but not to all older ones tested; they produced a toxin (called AF-toxin) that has the same biological specificity as that of the isolates. Unfortunately, the susceptibility (or sensitivity) of the cv. Morioka-16 to the disease (or AF-toxin) was inferred to be semidominant and controlled by a single gene pair (39). The interest was that the isolates (and the AF-toxin) can also be pathogenic (or toxic) to certain cultivars of Japanese pear, with the same host range as that of Japanese pear pathotype (AK-toxin producer) of *A. alternata*; AF-toxin appears to be chemically similar to AK-toxin (Nishimura *et al.*, unpublished). As is well-known, since about 1900 black spot of Japanese pear on susceptible Nijisseiki pear had caused predominantly by a single pathotype, AK-toxin producer, of *A. alternata*. Then, for about 80 years why did the Nijisseiki pear orchards remain free from another pathotype, AF-toxin producer ? Why are AF-toxin producers not detectable currently in pear orchards, so long as growers do not introduce the Morioka-16 cultivar of strawberry into the same cultivation areas as the pear orchards ? Studies of those interesting, but ecologically troublesome, implications with both types of *A. alternata* in the same susceptible host are in progress in our laboratories. The difference in the selection pressure involved here may be a difference of adaptation to its host between the two parasites.

IV. Concluding Remarks

Advances are now being made in determining the real roles of host-specific toxins as initiation factors of pathogenesis at the infection site of host - parasite interactions, but equal emphasis must be given to the roles of toxin as agents of selection pressure of the pathogen in agroecosystems.

Toxin studies in plant pathology are not merely of academic interest. However, our approach is still in its infancy, and the purpose of this chapter has been to stimulate thought and experiments rather to report on achievements.

In addition, the possibility for host-specific toxins and the closely related phytotoxins in practical application to agriculture are impressive. The application should be devised not only directly for the screening of resistant seeds or clones, but also indirectly in many ways. The future prospects for toxin studies in the areas of both basic and practical research are, therefore, truly exciting, interesting and challenging.

Acknowledgments

Thanks and appreciation are given to Dr. J. M. Daly for help in preparation of this review. We also wish to acknowledge many co-workers of *Alternaria alternata* toxin studies at Nagoya University, Tottori University and Kyoto University.

References

1. Bushnell, W. R., and Rowell, J. B., *Phytopathology 71*, lol2 (1981).
2. Comstock, J. C., and Scheffer, R. P., *Phytopathology 63*, 24 (1973).
3. Ellingboe, A. H., *in* "Active Defence Mechanisms in Plants" (R. K. S. Wood, ed.), p. 179. Plenum Press, New York and London, (1982).
4. Gnanamanikam, S. S., and Patil, S. S., *Phytopathology 66*, 290 (1977).
5. Hata, T., Sano, Y., Matsumae, A., Kamio, Y., Nomura, S., and Sugawara, R., *Japan J. Bacteriol. 15*, 1075 (1960).
6. Hayami, C., Otani, H., Nishimura, S., and Kohmoto, K., *J. Fac. Agr. Tottori Univ., 17*, 9 (1982).
7. Heath, M. C., *Phytopathology 71*, 1121 (1981).
8. Kohmoto, K., Khan, I. D., Renbutsu, Y., Taniguchi, T., and Nishimura, S., *Physiol. Plant Pathol. 8*, 141 (1976).
9. Leonard, K. J., *Plant Dis.Reptr. 56*, 834 (1972).
10. Leonard, K. J., *Phytopathology 63*, 112 (1973).
11. Leonard, K. J., *Plant Dis. Reptr. 60*, 245 (1976).
12. Matern, U., Beier, R., and Strobel, G. A., *Biochem. International 4*, 655 (1982).
13. Nakashima, T., Ueno, T., and Fukami, H., *Tetrahedron Lett. 23*, 4469 (1982).
14. Nishimura, S., *Proc. Japan Acad. 56 B*, 362 (1980).
15. Nishimura, S., Kohmoto, K., Kuwada, M., and Watanabw, M., *Ann. Phytopath. Soc. Japan 44*, 359 (1978).
16. Nishimura, S., Kohmoto, K., and Otani, H., *Rev. Plant Protec. 7*, 21 (1974).
17. Nishimura, S., Kohmoto, K., and Otani, H., *in* "Recognition and Specificity in Plant Host - Parasite Interactions" (J. M. Daly and I. Uritani, ed.), p. 133. Japan Scientific and University Park, Tokyo and Baltimore, (1979).
18. Nishimura, S., Kohmoto, K., Otani, H., Fukami., and Ueno, H., *in* "Biochemistry and Cytology of Plant - Parasite Interaction" (K. Tomiyama, J. M. Daly, I. Uritani, H. Oku and S. Ouchi, ed.), p. 94. Kosansha and Elsevier, Tokyo and Amsterdam, (1976).

19. Nishimura, S., Kohmoto, K., Otani, H., Ramachandran, P., and Tamura, F., *in* "Plant Infection: The Physiological and Biochemical Basis" (Y. Asada, W. R. Bushnell, S. Ouchi and C. P. Vance, ed.), p. 199. Japan Scientific and Springer-Verlag, Tokyo and Berlin, (1982).
20. Nishimura, S., Omura, T., and Kohmoto, K., *Ann. Phytopath. Soc. Japan 48,* 110 (1982).
21. Nishimura, S., and Scheffer, R. P., *Phytopathology 55,* 629 (1965).
22. Nishimura, S., Sugihara, M., Kohmoto, K., and Otani, H., *J. Fac. Agr. Tottori Univ. 13,* 1 (1978).
23. Nomura, S., Horiuchi, T., Hata, T., and Omura, S., *J. Antibiotics 25,* 365 (1972).
24. Nomura, S., Horiuchi, T., Omura, S., and Hata, T., *J. Biochem. 71,* 783 (1972).
25. Pdvody, G. N., Dunkle, L. D., and Edmunds, L. K., *Phytopathology 67,* 1485 (.977).
26. Omura, S., *Bacterial Rev. 40,* 681 (1976).
27. Otani, H., Nishimura, S., and Kohmoto, K., *Ann. Phytopath. Soc. Japan 40,* 59 (1974).
28. Otani, H., Nishimura, S., Kohmoto, K., Yano, K., and Seno, T., *Ann. Phytopath. Soc. Japan 41,* 467 (1975).
29. Pinkerton, F., and Strobel, G. A., *Proc. Natl. Acad. Sci. USA 73,* 4007 (1976).
30. Pringle, R. B., and Scheffer, R. P. *Ann Rev. Phytopath. 2,* 133 (1964).
31. Scheffer, R. P., *in* "Encyclopedia of Plant Physiology N. S. 4: Physiological Plant Pathology" (R. Heitefuss and P. H. Williams, ed.), p. 247. Springer-Verlag, Berlin, (1976).
32. Scheffer, R. P., and Pringle, R. B., *in* "Dynamic Role of Molecular Constituents in Plant - Parasite Interaction" (C. J. Mirocha and I. Uritani, ed.), p. 217. Bruce Publ. Co., St. Paul, (1967).
33. Scheffer, R. P., and Yoder. O. C., *in* "Phytotoxins in Plant Diseases" (R. K. S. Wood, A. Ballio and A. Graniti, ed.), p. 251. Academic Press, London, (1972).
34. Suzuki, S., Nishimura, S., and Kohmoto, K., *Ann. Phytopath. Soc. Japan 48,* 110 (1982).
35. Tamura, F., Nishimura, S., Kohmoto, K., and Otani, H., *J. Fac. Agr. Tottori Univ. 15,* 10 (1980).
36. Tsuge, N., and Nishimura, S., *Ann. Phytopath. Soc. Japan 48,* 360 (1982).
37. Ward, H. M., *Ann. Mycol. 1,* 132 (1903).
38. Wheeler, H., *in* "Toxins in Plant Disease" (R. D. Durbin, ed.), p. 477. Academic Press, London and New York, (1981).

39. Yamamoto, M., Nishimura, S., and Kohmoto, K., *Ann. Phytopath. Soc. Japan 49,* 110 (1983).

40. Yoder, O. C., *Ann. Rev. Phytopath. 18,* 103 (1980).

5

Future Prospects in Toxin Research

RICHARD D. DURBIN

I. Introduction

What of the future? I believe we are witnessing the beginning of an almost exponential increase in research and interest on microbial toxins and the role they play in disease causation. This has come about because of several developments: i) It has become increasingly evident that toxins are important disease determinants and, while they are not always required for pathogenesis, their synthesis is broadly distributed among pathogenic fungi and bacteria. In addition, the symptoms produced by other kinds of pathogens strongly suggest that they too utilize toxins in pathogenesis. ii) Information about the chemistry of toxins has caught up with and, in some cases, surpassed what biological information we possess. This fact should allow us to "leap forward" our biological understanding based on the use of purified, structurally defined compounds whose fate can be followed quantitatively. iii) The application of new and emerging technologies, developed for molecular biology, to toxin studies has the potential to greatly extend our understanding of toxigenesis. iv) Various exceeding sensitive and/or specific methods for detecting compounds and their interactions have been developed which can be applied to toxins. v) Finally, because toxins are relatively small molecules (generally less than 1,000 daltons) and have defined mechanisms of action, they lend

TOXINS AND PLANT PATHOGENESIS
ISBN 0 12 200780 8

themselves well to theoretical and practical studies on the nature of disease resistance, host specificity, etc.

Truly, it seems we are on the threshold of being able to answer many fundamental questions at the molecular level which in time should be reflected in practical applications. My purpose here is to focus attention on what these questions might be, to point out some areas for fruitful investigations, and, hopefully, to raise some provocative possibilities.

II. Biochemical Considerations

A. *Structure*

The recent advances made in elucidating toxin structures, especially of the host-specific fungal toxins, is most heartening, and paves the way towards a meaningful solution of what their sites of action are and how the toxin-target interactions function. Much of this work has been made possible by the powerful analytical instrumentation which has been developed to facilitate structural analysis (*e.g.*, GC-MS, ^{1}H and ^{13}C NMR, and X-ray analysis). The importance of this structural information, and the attendant availability of verifiably pure compounds, can hardly be over emphasized for they provide the "keys" which will allow us to exploit many other areas of toxin research: for instance, biosynthesis, metabolic regulation, mechanism of action, structure-activity, and metabolism.

There has been an increasing shift to a team approach for structural determinations. This *modus operandi* needs to be encouraged even more, for the single scientist, "jack of all trades" approach so many of us have had to use in the past is simply not adequate. The long-term success of such teams lie not only in bringing together the required disciplines, but, most importantly, in an ability of the members to have some comprehension of the others' areas of expertise. This will necessitate modifying existing curricula somewhat to provide even more physical science and mathematical training for the biologists and more biological training for the physical scientists.

While the total number of examples is still relatively small, there has been increasing activity on the chemical synthesis of toxins (Rich, 1981). For some, the raison d' être has been to unequivocally establish structure; for others, the purpose has been to allow study of the relationship between structure and biological activity, or as a method for producing high specific activity, specifically labeled toxins. These latter two purposes will surely provide a growing stimulus for devising synthetic reactions and routes for more toxins as we move increasingly into mechanism of action and metabolism studies. In addition to these reasons, it may more advantageous to produce some toxins synthetically or semi-synthetically, than to isolate them from culture filtrates.

One of the more serious constraints we face is the lack of ready sources of pure toxins. Each of us working with a particular system tends to take for granted both the art and science required for the uninitiated to produce and purify the toxins we may have spent years learning about. This task is even more daunting to those in other disciplines more interested in simply using toxins as tools, for generally they do not have expertise in microbiology or natural product isolation. In the case of tentoxin, it has been cost effective for a commercial biochemical supply company to sell it. Not withstanding this, I know of several investigators who, because of cost, have been forced to drop research on this toxin. What the solution is, I do not know. But it is a problem we need to seriously address and hopefully solve. Baring this, future advancement will be much slower and sporatic.

B. *Biosynthesis and its Regulation*

Biosynthetic studies have and are being done on a variety of toxins. However, paradoxically, there has been relatively few such studies on those toxins for which there is the most evidence for involvement in disease causation. In part, this is because we still do not have sufficient structural information on these compounds. I suspect it is also because this kind of metabolic study has not attracted the interest of many of the appropriate specialists. The problem is further aggravated by the fact that the *in vitro* yields of many toxins are generally quite low, and, unfortunately, little effort has been made to increase them by either genetical or cultural means. The striking increases in toxin production obtained when plant cells (Larkin and Scowcroft, 1981) or intercellular fluids (Durbin, unpublished data) are used as substrates is but one illustration of our ignorance of how toxin production is controlled. We are also confronted, in some instances, with the problem of maintaining the toxigenic capacity of cultures over time. This is a problem area which could benefit greatly from an increased genetical input.

As part of the biosynthetic studies on toxins, it would also be useful to determine their site of synthesis within the microbial cell. It seems likely that this synthetic machinery might be membrane-bound, and thus compartmentalized from its target. If so, this information would have relevance to studies on toxin translocation and mechanisms for self-protection (see succeeding sections).

The use of specifically labeled ^{13}C-containing precursors needs to be expanded, for stable isotope NMR spectroscopy is an exceedingly powerful analytical tool for structural analysis that can quickly, definitively and nondestructively yield results. Even now whole-cell, real-time determinations of metabolic transformations are within our grasp [*e.g.*, energy change, pH, labeled substrates and pyridine nucleotides (Unkefer *et al.*, 1983)]. Certainly the contribution of NMR spectroscopy to our understanding of how toxins work will expand as spectroscopic specialists enter the field and suitable instrumental probes are developed.

It is useful to remind ourselves that toxins are typically products of secondary metabolism. For instance: i) each is produced by a restricted number of taxonomically related pathogens, ii) they are produced at a specific phase in the pathogen's life cycle, iii) their production is highly dependent upon nutritional and environmental conditions, iv) many toxins are the result of complex and unusual biochemical transformations, v) they have no obvious function in the maintenance of the pathogen's primary processes, and vi) they are not readily metabolized. Considering this, it therefore seems probable that toxin biosynthesis is controlled by the same kinds of regulatory processes that control the production of other secondary metabolites.

From this viewpoint, the study of toxin biosynthesis takes on an added dimension, for an understanding of how and where regulatory control is exercised could have important practical applications. Special attention should be paid to the first reaction unique to the toxin itself. For by analogy with similiar kinds of pathways, it is this step which generally is pivotal and holds the key to controlling flux through the remainder of the pathway.

If, for example, it can be shown that the production of a toxin is under metabolic control, it suggests the possibility that we might be able to manipulate the environmental conditions or host so as to modulate the toxin's synthesis. For instance, conditions for its production in culture could be maximized. Conversely, in the host, one might be able to "switch it off" or significantly reduce its synthesis. A possible example of this is contained in a recent study on the aflatoxin content of various maize varieties and hybrids which showed that one of the factors responsible for the observed difference in aflatoxin content is genetic (Zuber *et al.*, 1983). If this factor is of major importance, it would be feasible to develop maize genotypes having genetic control of aflatoxin formation. Conceivably, this control over synthesis might sometimes operate at the regulatory level.

Some support for the notion that toxin production can be regulated by chemical cues from the host comes from studies on *Helminthosporium sacchari*. Tox$^-$ strains of this fungus have been reported to resume toxin production when serinol (2-amino-1,3-propanediol) is added to the culture medium and lower concentrations of this compound appear to be present in resistant cultivars (Babczinski *et al.*, 1978). However, other workers found that co-culturing the fungus with sugar cane cell suspensions gave much higher and earlier toxin production than did serinol (Larkin and Scowcroft, 1981). But, no differential effect on HS-toxin production between susceptible and resistant cultivars was observed.

Comparable studies on bacterial pathogens, suggesting that there may be natural modulators of toxin production, have not been reported; however observations on several pathogens are suggestive of this phenomenon. In the cases of *Pseudomonas syringae* pv. *phaseolicola* (Gnanamanickman and Patil, 1976) and *Rhizobium japonicum* (Owens and Wright, 1965), selected cultivars will not support toxin production even though the bacterium multiplies to a level that would have resulted in toxin production in other cultivars. While studying the production of tabtoxin by *P. syringae* pv. *tabaci*, we found that *in vitro* the intercellular fluid from tobacco would immediately cause tabtoxin production to stop and tabtoxinine-ß-lactam, the biological active form of the toxin, to begin accumulating. That the fluid may also control the form of the toxin produced *in vivo* is indicated by our ability to isolate only tabtoxinine-ß-lactam from infected plants. Another example of the phenomenon involves *P. syringae* pv. *tagetis*. In our studies on this pathogen, we have collected 14 isolates from various hosts in different parts of the world. Of these only one isolate, that from Australia, will produce the toxin in culture; the rest of the isolates require the host for toxin production.

C. *Mechanism of Action*

At present, we do not have a complete understanding for the mechanism of action of any toxin. This should not be disturbing, given the state of the art and the degree of complexity involved. With regards to this "state of the art," toxin researchers have increasingly found it necessary to become primary providers of biochemical and physiological knowledge. Only after such information has been provided, have they been able to proceed in studying toxin action (*e.g.*, fusicoccin and tentoxin). Although this state of affairs makes for difficulties and slower progress, it is a good sign in the sense that the work in question is on the forefront of biological experimentation and thought. It also means we have the potential for making contributions to a broader audience.

As the number of toxins being studied increases, a broader range of kinds of targets can be expected. At the present time several organellar targets have been identified (*e.g.*, chloroplasts, mitochondria, plasmalemma). This is in contrast to the molecular level where enzymes are the only targets which have been identified. Scientists have described several kinds of enzyme inhibition kinetics but even so these probably represent only a portion of the kinds of toxin-target interactions there are. In particular, elucidation of how toxins interact with membranes and membrane-bound enzymes should provide novel information. Are there perhaps analogies with animal peptide hormones?

How are binding events transduced into biochemical and/or electrochemical events? Possibly, they lead to detectable perturbations in membrane structure. If so, critical characteristics such as membrane fluidity could be examined by a variety of techniques, for instance electron spin resonance spectroscopy, fluorescent anisotropy, and whole-cell NMR. Photo affinity probes are available which will label proteins when a toxin inserts into the membrane's hydrophobic regions (Thelestam *et al.*, 1983).

Until now we have not examined in detail many different toxins which induce either necrosis, chlorosis, watersoaking, or growth abnormalities. It will be interested to see whether, as we acquire more information, we will find a predominance of molecules with specific properties (*e.g.*, hydrophobicity, molecular size, or specific functional groups) or actions (*e.g.*, inhibitors of enzymes, membrane function, or transcription-translation processes) associated with a particular host response.

D. *Specificity*

The acquisition of a body of definitive structural information on host-specific toxins will finally allow us to examine the basis for their specificity in a meaningful way. Not that this information *per se* will directly provide the answers, but rather its attainment means we can begin to label and quantify pure compounds. We already know that these toxins are not all structurally similiar, which makes it quite likely that the biochemical explanation for their specificity will also vary. Likewise, their structures do not immediately suggest why they are host-specific. Host cells may be replete with structural features which potentially could serve as receptors. The pathogen, with its plastic capacity to synthesize a broad range of small molecules may be simply exploiting the situation. In the evolution of parasitism, we may indeed constantly have the production of new "toxins" and "targets," both in search of one another!

The ability to produce high specific-activity radiolabeled toxin probes will be particularly crucial for understanding host specificity, as it appears in many cases that the number of toxin binding sites in the host is limited. The observation that the double bond of the O-t-pentenyl group on the glucose moiety of fusicoccin (FC) can be reduced without loss of toxicity has played a critical role in allowing rapid progress in mechanism of action studies on this toxin. Carrying out the reduction with ^{3}H has yielded dihydro-FC with a specific activity as high as 62.5 Ci mmole^{-1} (Ballio *et al.*, 1978). Even with this kind of probe we may approach the maximum possible detection limits of an intact system. However, dissection of the system and enrichment of the target will be fraught with problems (*e.g.*, denaturation and degradative enzymes acting on the toxin). Generally, we will be limited in size of the target sample so that additional techniques may need to be employed to amplify the toxin signal and enhance the signal to noise ratio. Coupling toxins to horseradish peroxidase for visualization purposes would be an example of this (Chester *et al.*, 1979). Likewise, a fluorescent label is capable of four times the optical resolution of a ^{3}H label. Some of the techniques currently being used in nucleic acid hydridization and immunoradiometric assays also may be very useful for *in situ* and *in vitro* localization and detection of toxin-target interactions. Their use will, however, require appropriate groupings on the toxin to attach chemical "linkers" which can in turn be joined to immunological, fluorescent, or affinity reagents (Bayley and Knowles, 1977; Chowdhry and Westheimer, 1979).

Radioiodination, a technique commonly employed in other disciplines to study receptor binding, has not been applied to toxins, since they either lack tyrosyl residues or because the introduction of ^{125}I would change the nucleophilicity of critical amino groups in the toxin. With toxins of low molecular weight, this alteration would probably abolish their biological activity. Recently however, a new method for derivatization and radioiodination, has been developed which conserves nucleophilicity (Su and Jeng, 1983). This technique may be suitable for some toxins, if the reaction does not sterically affect target accessibility.

E. *Movement and Metabolism in Host*

Once produced, how does a toxin move from its site of synthesis in the pathogen across the barriers presented by one or more membranes and cell walls to its target in the host? A particularly interesting question in this regard is whether an inability to traverse these barriers in the host could be the basis of a resistance mechanism.

Staskawicz and Panopoulos (1980a) have shown that phaseolotoxin can take advantage of an oligopeptide transport system to gain "illicit" entry by virtue of its tripeptide portion, ornithylalanylhomoarginine. This tripeptide does not, however, affect the toxin's interaction with its target, ornithine carbamoyltransferase. One wonders if this portion of the molecule has not evolved specifically to facilitate translocation.

Related to this phenomenon is the work of Carlson (1973), who isolated several mutant tobacco lines which were resistant to chlorosis induction by *Pseudomonas syringae* pv. *tabaci*. According to Carlson, the basis for this resistance was the high levels of methionine contained in the cells of these lines. Since methionine is known to compete with methionine sulfoximine (MSO) for transfer across a membrane (Meins and Abrams, 1972) and MSO acts very similiarly to tabtoxin in plants (Frantz *et al.*, 1982), he postulated that the methionine was antagonizing the entry of tabtoxin into the host cell.

From all these results, we can infer that specific portions, or functional groups, may be needed for toxin transport across membranes and, furthermore, that this movement can be antagonized. Accordingly, we should then be able to identify specific genotypes of the host which when expressed would inhibit toxin entry.

Toxin transformations in the host are poorly understood. Once again, this stems from our lack of knowledge of toxin structure and inability to produce high specific-activity labeled toxins. Metabolic alterations may be very important for toxin activation, as for example in the case of tabtoxin where host and/or bacterial enzyme action is required to convert tabtoxin to the biologically active tabtoxinine-ß-lactam form (Uchytil and Durbin, 1980). Likewise, toxins may be degraded into less or nontoxic forms [*e.g.*, similar to the degradation of HS-toxin by ß-galactofuranosidase (Livingston and Scheffer, 1982)]. Both kinds of transformations are potentially important considerations in terms of host resistance, and we need to know how widespread post-synthesis modifications are. On the one hand, a plant be able to block activation or, on the other, promote inactivation and degradation.

F. *Symptom Production*

Very little is known about the physico-chemical events linking the primary toxin-target interaction to the resulting symptom(s). An understanding of this linkage would have value for several reasons besides its intrinsic interest. Knowledge of the events in the linkage offers us the potential opportunity to obviate the symptom, and perhaps control the disease, by altering the course of one or more reactions in the linkage. In addition, it would represent a validation of the primary mechanism of action proposed for the toxin. By this I mean that even when a specific toxin-target interaction has already been identified, the crucial question is whether the biochemical consequences of this interaction can fully explain the effects observed on the host. In reality, the toxin might have multiple sites of action, and perhaps the site initially proposed may not be metabolically linked to the symptom observed. Actually, we have an increasing number of reports that the same toxin may have different mechanisms of action: T-toxin (Daly, 1981; Earle and Gracen, 1981), tabtoxin (Crosthwaite and Sheen, 1979; Thomas *et al.*, 1983), phaseolotoxin (Patil *et al.*, 1970; Smith and Rubery, 1982), tentoxin (Steele *et al.*, 1976, Duke *et al.*, 1982), and rhizobitoxine (Giovanelli *et al.*, 1971; Lieberman, 1979).

A presumption based upon the commonly held definitions of a toxin and how we test for their presence is that they must produce a visual response in the host. Is this valid? Could not toxins act in some manner which simply allows another attack mechanism to function? The definitions also imply that toxins act independently rather than in concert with other disease determinants in a parallel yet coordinated and/or sequential fashion. We need to question these implications, and to test whether and how toxins might interact with other disease determinants. We are especially hampered in this quest by current thought and practice. How many toxins might we be missing because we have oversimplified the bioassay system, or not provided a sufficiently complex and dynamic environment for toxin production? I am intrigued by the sequence of events postulated by Mussell and his colleagues to occur in Verticillium wilt (Mussell and Stilwell, 1982): endopolygalacturonase from the fungus releases host cell wall-bound mycelial degrading enzymes which then solubilize from the pathogen substances toxic to the host. Only by using a complex bioassay system would this work have been feasible.

G. *Self-protection Mechanisms*

In some instances, the toxin's site of action is present in the pathogen as well as in the host. How then does the pathogen avoid suicide? What kinds of mechanism for self-protection does it possess? Perhaps, if we knew what the bases of self protection are, we could, in selected instances, develop some method for upsetting this mechanism--in essence force the suicide!

Miller and Brenchley (1981) have reported that the basis for resistance to MSO among some strains of *Salmonella typhimurium* lies in the response of the target, glutamine synthetase, to MSO. The resistant forms had significant increases in apparent Km's for glutamate and ammonia and lack transferase activity. Although it is not clear that these findings can wholly explain resistance, they do illustrate the possibility that studies of self-protection and resistance mechanisms in microorganisms might provide some useful models for understanding higher plant resistance.

III. Biological Considerations

A. *Function*

I have argued that the evolutionary conservation of genetic information encoding for toxin synthesis strongly supports the argument that toxin production represents an evolutionary event which provides a selective survival advantage (Durbin, 1982, 1983). If its production had no positive advantage for the pathogen, one would not expect to find a high proportion of the pathogen population with this character. Rather, it would tend to either drop out or be found at only low frequencies in the population. This indeed may be the case with rhizobitoxine (Owen and Wright, 1965). In instances where the toxin's target is also present in the pathogen, there is an even greater probability that toxin production has some positive function, since additional genetic information must be utilized to protect the pathogen from its own toxin. Clearly, both production and self-protection represent significant expenditures of metabolic energy.

The production of tabtoxin, tentoxin, and coronatine, each by groups of closely related species, is most readily viewed as adaptive radiation from a single ancestral species. In addition, these and other toxins affect a wide range of organisms; however some organisms exist that are resistant. It is difficult to rationalize these facts without there being some commensurate benefit to the pathogen.

While this argument appears to be self-evident for those toxins which have been shown to be primary disease determinants, there are many examples of toxins presumably acting as secondary disease determinants for which no advantage to the pathogen has yet been found. Perhaps some of these toxins have essentially no value to the pathogen during parasitism. Certainly the production of natural products harmful to plants does not in itself establish that these products (*i.e.*, toxins) are produced as an evolutionary response to parasitism. The fact that a toxin affects a plant may be purely a matter of chance. Apropos of this possibility, Turner (1971) has pointed out that the biosynthetic origins of secondary metabolites with and without antibiotic activity "are often so similar that the possession of antibiotic activity must be regarded as a fortuitous property of the product."

We need to take a wider prospective. For instance, the answer instead may lie in considering the saprophytic phase of the pathogen (*i.e.*, as it occurs in the soil, decaying organic matter, or as an epiphyte in the phylloplane or rhizoplane). Could not toxin production be related to resource competition? Janzen (1977) has made the provocative suggestion that one of their functions might be to make decaying fruits less palatable to insects or mammals, who otherwise might consume the substrate. Viewed in this light, some toxins might more appropriately be classified as antibiotics. This is essentially the case with a socalled antibiotic produced by a *Streptomyces* sp. which differs from tabtoxinine-ß-lactam only having a chlorine instead of a hydrogen at C-4 (Scannell *et al.*, 1975).

Some "toxins" may be nothing more than cultural artifacts. About 30 years ago Dimond and Waggoner (1953) promulgated criteria for establishing that a toxin functions in disease production. Since then, others have refined and extended the criteria, and created additional categories (Graniti, 1972; Wheeler and Luke, 1963; Yoder, 1980). However, the basic goal of all these systems is the same: to precisely understand what role the compound has in disease causation. Unfortunately, because the criteria for proving complicity are difficult to fulfill, the number of toxins for which we have convincing evidence for their participation in pathogenesis is still quite small. Nevertheless, more effort needs to be made in this direction. For we still continue to see too many reports of compounds purported to be "phytotoxins" for which even presumptive evidence for complicity in disease causation is tenuous.

B. *Resistance*

It perhaps goes without saying that understanding the basis for toxin resistance is one of our major goals. Elucidating this goal will bring to bear many of the research areas about which I have already spoken. Accordingly, we should not focus attention on it to the exclusion of the other areas, albeit an exercise of which we have been guilty. My personal feeling is that we are more likely to serendipitously illuminate potential resistance mechanisms while examining any number of other questions than by studies specifically addressing resistance.

An old hypothesis is that plants may be resistant to toxins because they degrade them. Still another hypothesis which needs to be considered is that the toxin activates a self-repair mechanism in resistant plants (Wheeler, 1981). While these are attractive possibilities, there is no very substantial evidence as yet that such mechanisms operate, let alone are widespread. Nevertheless, we should not abandon these ideas for only a very few systems have been thoroughly examined. An alternate viewpoint states that the target is absent or its binding sites are unavailable for interaction with the toxin. For this explanation we do have some examples: phaseolotoxin (Staskawicz *et al.*, 1980b) and tentoxin (Steele *et al.*, 1976). A modification of this viewpoint is that the target is present but has an altered affinity for the toxin. Model systems can be constructed which show that rather subtle changes in kinetic parameters can have profound effects on toxin-target interactions.

Potentially, a host could possess multiple mechanisms of resistance against any one toxin. This notion needs to be considered when developing experimental procedures to screen for toxin resistance. Otherwise, the full range of resistant host genotypes may not be detected.

C. *Genetics*

With regards to toxin production, we know essentially nothing about primary gene structure, how gene expression is regulated, or how the genome is organized. Fungi with sexual stages somewhat aside, future work will be strongly influenced by the tools of molecular genetics which can be applied to a particular case. Whereas there are already a multitude of potential transformation methods and cloning vehicles for use with procaryotes, the availability of comparable systems for use with toxigenic fungi is still very limited. Some of the technologies used in yeast studies, such as autonomously replicating sequences and 2μ circles, an endogenous plasmid, may be more generally applicable as eucaryotic cloning vehicles. They need to be examined (Hinnen, 1978; Chan, 1980). Genes for the production of many toxins conveniently lend themselves for use as selectable transformation markers, so that if problems associated with the rate of transformation and stable inheritance of cloned DNA can be overcome, major advances should be forthcoming.

The current NIH "Guidelines for Research Involving Recombinant DNA Molecules" (1982) have not placed any phytopathogenic organisms on the restricted use list nor do they put any specific restrictions on cloning experiments involving toxins with a vertebrate LD_{50} of 100 μg or more per Kg body weight. Various containment conditions are specified for toxins with toxicities between 100 μg and 100 ng; cloning of genes coding for toxins with an LD_{50} of 100 ng or less is still prohibited. Although very little data is currently available on the vertebrate toxicity of most toxins, other than those compounds more commonly considered as mycotoxins, these restrictions do not appear to be overly restrictive for those workers observing the guidelines--tabtoxin, for example, requires about 150 μg/kg to produce convulsions without death (Sinden *et al.*, 1968). In addition, many phytopathogenic bacteria have resident plasmids or other mobilizing systems which allow for genetic interchange, and hence remove these bacteria from consideration under the Guidelines. Thus, as training in and appreciation of the potential use of molecular genetic becomes more widespread, we should see their rapid application to toxin studies. Furthermore, we can expect that future progress in genetic engineering will facilitate the study of toxins even more.

D. *Taxonomic Distribution*

As the number of examples of toxins from diverse taxonomic groups increases, we should expect more toxins with novel structural features and mechanisms of action to be identified. In this regard, with the exception of *Spiroplasma citri* (Daniels and Meddins, 1974), we still have essentially no information about the production or significance of toxins in diseases caused by spiroplasmas, mycoplasmas, or rickettsias. The same can be said of nematodes. Our inability to readily culture these pathogens on chemically defined substrates is mainly responsible for this state of affairs, and not until we can do this will there be appreciable progress.

E. *Applications*

As we learn more about the properties of toxins and how they act, we need to consider how they might be used in other contexts. Because of their specificity or physical properties, some have already found use as biochemical probes, in commerce, and for the screening of plant material for toxin resistance (Durbin, 1981). These properties insure that there will be an increased use of these substances in a wide variety of disciplines and problems. We need only to elucidate them.

For some years workers have visualized the possibility of screening host materials for toxin resistance at the single cell level, and indeed this approach has proven to be successful in selected cases (Durbin, 1981; Earle and Gracen, 1981). It is interesting that, where appropriate results are available, the number of resistant lines obtained is similar whether or not a mutagenic treatment is applied. If the natural response to a toxin does vary among cells within a single plant, it would certainly make the task of obtaining resistant cells much easier in species where haploid genomes are unavailable.

Barton and Brill (1983) have recently suggested that molecular biology will eventually play a major role in developing resistance to plant pathogens. This seems likely. Certainly, the possibility of developing toxin-resistant plant materials has not gone unnoticed in the marketplace (Chaleff, 1983). University and Government scientists must realize this and come to an accord with their industrial counterparts. Otherwise, if a problem is of sufficient economic importance, the former will be overwhelmed.

IV. Conclusions

Although the recent past has provided many major contributions to our understanding of toxins and their action in plants, there still remains many significant areas about which we have little knowledge. It is apparent that future advances will depend in large measure upon a consortium of scientists from diverse disciplines working together, and upon advances being made in seemingly unrelated areas, particularly those concerned with analytical technologies.

Generally, our researches on toxins soon turn to examining them and their interactions as isolated, *in vitro* phenomena. This is as it should be. However, once we have defined this state, we must again move back into the environment of the diseased plant if we are to fully understand how toxins fit into the dynamic web of biochemical events constituting pathogenesis.

References

Anon. (1982). *Fed. Reg. 47*, 38048.
Babczinski, P., Matern, V., and Strobel, G. A. (1978). *Plant Physiol. 61*, 46.
Ballio, A., Federico, R., Pessi, A., and Scalorbi, D. (1980). *Plant Sci. Lett. 18*, 39.
Barton, K. A., and Brill, W. J. (1983). *Science 219*, 671.
Bayley, H., and Knowles, J. R. (1977). *Methods Enzymol. 46*, 69.
Carlson, P. S. (1973). *Science 180*, 1366.
Chaleff, R. S. (1983). *Science 219*, 676.
Chan, C. S., and Tye, B. (1980). *Proc. Natl. Acad. Sci. 77*, 6329.
Chester, J., Lentz, T. L., Marquis, J. K., and Mautner, H. G. (1979). *Proc. Natl. Acad. Sci. 76*, 3542.
Chowdhry, V., and Westheimer, F. H. (1979). *Ann. Rev. Biochem. 48*, 293.
Crosthwaite, L. M., and Sheen, S. J. (1979). *Phytopathology 69*, 376.
Daly, J. M. (1981). *In* "Toxins in Plant Disease" (R. D. Durbin, ed.), pp. 331. Academic Press, New York,
Daniels, M. J., and Meddins, B. M. (1974). *Collog. INSERM 33*, 195.
Dimond, A. E., and Waggoner, P. E. (1953). *Phytopathology 43*, 229.
Duke, S. O., Wickliff, J. L., Vaughn, K. C., and Paul, R. N. (1982). *Physiol. Plant. 56*, 387.
Durbin, R. D. (1981). *In* "Toxins in Plant Disease" (R. D. Durbin, ed.), pp. 495. Academic Press, New York.
Durbin, R. D. (1982). *In* "Phytopathogenic Prokaryotes," (M. S. Mount, and G. H. Lacy, eds.), Vol. I, pp. 423. Academic Press, New York.
Durbin, R. D. (1983). *In* "Biochemical Plant Pathology" (J. A. Callow, ed.), p. 137. John Wiley & Sons, Inc., New York.
Earle, E. D., and Gracen, V. E. (1981). *In* "Plant Disease Control" (R. C. Staples, and G. H. Toenniessen, eds.), p. 285. John Wiley & Sons, New York.
Frantz, T. A., Peterson, D. M., and Durbin, R. D. (1982). *Plant Physiology 69*, 345.
Giovanelli, J., Owens, L. D., and Mudd, H. (1971). *Biochem. Biophys. Acta 227*, 671.
Gnanamanickman, S. S., and Patil, S. S. (1976). *Phytopathology 66*, 290.

Graniti, A. (1972). *In* "Phytotoxins in plant diseases" (R. K. S. Wood, A. Ballio, and A. Graniti, eds.), p. 1. Academic Press, London, England.

Hinnen, A., Hicks, J. B., and Fink, G. R. (1978). *Proc. Natl. Acad. Sci. 75*, 1929.

Janzen, D. H. (1977). *Amer. Nat. 111*, 691.

Larkin, P. J., and Scowcroft, W. R. (1981). *Plant Physiol. 67*, 408.

Lieberman, M. (1979). *Ann. Rev. Plant Physiol. 30*, 533.

Livingston, R. S., and Scheffer, R. P. (1982). *Phytopathology 72*, 933.

Meins, Jr., F., and Abrams, M. L. (1972). *Biochem. Biophys. Acta 266*, 307.

Miller, E. S., and Brenchley, J. E. (1981). *J. Biol. Chem. 256*, 11307.

Mussell, H., and Stilwell, P. (1982). *Phytopathology 72*, 264.

Owens, L. D., and Wright, D. A. (1965). *Plant Physiol. 40*, 927.

Patil, S. S., Kolattukudy, P. E., and Dimond, A. E. (1970). *Plant Physiol. 46*, 752.

Rich, D. H. (1981). *In* "Toxins in Plant Disease" (R. D. Durbin, ed.), pp. 295. Academic Press, New York.

Scannell, J. P., Pruess, D., Blount, J. F., Ax, H. A., Kellett, M., Weiss, F., Demny, T. C., Williams, T. H., and Stempel, A. (1975). *J. Antibiotics 28*, 1.

Sinden, S. L., Durbin, R. D., Uchytil, T. F., and Lamar, Jr., D. (1968). *Toxicology and Appl. Pharm. 14*, 82.

Smith, A. G., and Rubery, P. H. (1982). *Plant Physiol. 70*, 932.

Staskawicz, B. J., and Panopoulos, N. J. (1980a). *J. Bacteriol. 142*, 474.

Staskawicz, B. J., Panopoulos, N. J., and Hoogenroad, N. J. (1980b). *J. Bacteriol. 142*, 720.

Steele, J. A., Uchytil, T. F., Durbin, R. D., Bhatnagar, P., and Rich, D. H. (1976). *Proc. Natl. Acad. Sci. 73*, 2245.

Su, S., and Jeng, I. (1983). *Anal. Biochem. 128*, 405.

Thelestam, M., Jolivet-Reynaud, C., and Alouf, J. E. (1983). *Biochem. Biophys. Res. Comm. 111*, 444.

Thomas, M. D., Langston-Unkefer, P. J., Uchytil, T. F., and Durbin, R. D. (1983). *Plant Physiol.* (in press).

Turner, W. B. (1971). "Fungal metabolites." Academic Press, London.

Uchytil, T. F., and Durbin, R. D. (1980). *Experimentia 36*, 301.

Unkefer, C. J., Blazer, R. M., and London, R. E. (1983). *Science*. (In press).

Wheeler, H. (1981). *In* "Toxins in Plant Disease" (R. D. Durbin, ed.), p. 477. Academic Press, New York.

Wheeler, H., and Luke, H. H. (1963). *Ann. Rev. Microbiol.* *17*, 223.

Yoder, O. C. (1980). *Ann. Rev. Phytopathol.* *18*, 103.

Zuber, M. S., Darrah, L. L., Lillehoj, E. B., Josephson, L. M., Manwiller, A., Scott, G. E., Gudauskas, R. T., Horner, E. S., Widstrom, N. W., Thompson, D. L., Bockholt, A. J., and Brewbaker, J. L. (1983). *Plant Disease 67*, 185.

Index

Numbers refer to pages; *f* to figures; *t* to tables

3 4 5 6 7 8 9 0 1 2
A B C D E F G H I J